Studies On
Rice Collar Rot in Manipur

Studies On Rice Collar Rot in Manipur

L. N. K. SINGH

RAJESH PUBLICATIONS
NEW DELHI

Ist Published in India : 1997
ISBN. 81-85891-10-9

Published by :
M. L. GUPTA
Rajesh Publications
1, Ansari Road, Daryaganj
New Delhi-110002

Typesetting by :
ANITA GUPTA
Sanjay Enterprises
New Delhi-110002

Printed in India at :
Aditya Offset Press, New Delhi - 110 002

TO MY PARENTS

WITH

AFFECTION AND GRATITUDE

Preface

Rice is a staple food for about half the population of the globe. It is mostly consumed by the people in the countries of eastern hemisphere. Such important crop is often damaged by varieties of micro-organisms. Scientists all over the world have undertaken researches on the diseases of rice, and sufficient literature on different types of diseases of rice are available. However, the damages caused by the collar rot diseases have never been throughly analysed. So far four fungal pathogen actively involved in the collar rot disease have been recorded. However, I have been observing certain symptoms developed in the rice plant somewhat similar to that of collar rot disease, and hence, my main focus was to find out other causal organisms involved in the disease.

After intensive studies and researches I detacted and isolated two new fungal organisms actively involved in the collar rot disease. These findings, I am confident, may give a new vista to the scholars engaged in study of rice disease.

While attempting to find out these new micro organisms I had the opportunity to consult experts and scientists in and outside the state and abroad including the scientists and scholars of I.A.R.I., New Delhi. I had the advantage of consulting libraries of I.A.R.I., Reginoal Research Laboratory (R.R.L.) Jorhat. Assam Agricultural University, Jorhat, National Library, Calcutta, Manipur University, Regional Institute of Medical Sciences, Imphal and Central Agricultural University, Imphal.

I am immensely thankful to Dr. N. Iboton Singh, Associate Professor and Head, Department of Plant Pathology College of Agriculture , Central Agriculture University, Imphal for his supervision, suggestions and help during the course of my research work. I am also grateful to Dr. R.K. Tombisana Devi, Assistant Professor in the same Department for valuable suggestions, to Professor M. P. Singh, Vice-Chancellor and Professor S. Bapu Singh, Dean of the College for their valuable help in availing laboratory and library facilities. I also acknowledge the help rendered by Dr. M. Premjit Singh, L. Gourabidhu

Singh, S. Jeken Singh and N. Somorjit Singh of the College of Agriculture during the course of my study.

The author is grateful to Prof. A. K. Chatterjee, Dean, School of Science, and Head of Life Sciences Department Manipur University, and Dr. P. N. Chowdhary of IARI, New Delhi for their valuable help during the course of investigation. I also thank my fellow research scholars S. Kananbala Devi, W. Indira Devi, Y. Sushila Devi, R. K. Premabati Devi, Samjay Singh, S. Basanta Singh, Dr. Ch. Sarojnalini Devi, Research Associate and Dr. P. Kumar Singh, Assistant Professor in the Department of Life Sciences, Manipur University for their help and suggestions.

The author is grateful to his father Professor L. Janmejay Singh of Department of Life Sciences, Manipur University, his mother, brothers, sister and sister-in-law, and Mrs. N. Ibema Devi for their help and inspiration during the course of his study. I gratefully acknowledge the financial assistance provided to me in the form of Junior Research Fellowship by the Directorate of Science and Technology, Government of Manipur, which helped me in completing this study.

Lastly, I am thankful to Shri M. L. Gupta of Rajesh Publications, New Delhi for timely publication of this work.

It will give me immense satisfaction if this work provides some information to the researchers and scholars working in this field. Useful suggestions and healthy criticism to improve this book will be welcomed and thankfully acknowledged.

Khwai Naga Mapal
Kangajabi
Laikangbam Leikai
Imphal - 795004

L. Nongdren Khomba Singh

Contents

List of Tables

Chapter 1

Introduction

Rice (*Oryza sativa* L.) is the staple food of the people of Manipur. It occupies large cultivable areas in the state. Rice suffers from various diseases (Ou, 1985) resulting in heavy loss every year. Among the diseases, collar rot has become increasingly important in valley areas of Manipur.

Collar rot of rice caused by four species of *Ascochyta* was first reported by Hara, (1959). The disease is reported to be caused by *Ascochyta oryzae* Catt. from Thailand, (Kanjanasoon, 1962); South East Asia (Punithalingam and Sharma, 1979) and Manipur (Singh and Devi, 1991). However, collar rot is also reported to be caused by *Pestalotiopsis versicolor* (Speg.) from Manipur (Singh *et al*., 1995) and by *Pyricularia oryzae* from Malays (Chin, 1974). Not only these fungi mentioned above recently, *Chaetomium globosum* which can cause rice collar rot is also identified in Manipur.

Ascochyta oryzae produces abundant pycnidia on sterilized rice stems (Kanjanasoon, 1963). Wongsasitorn & Supavita (1974) studied varietal reaction of rice collar rot and host range of *A. oryzae*.

Very little is known about the ecological relationship, if any, of the fungi associated with rice collar rot, the physiology of the fungus, control aspect of the disease, etc. in India particularly in Manipur. The present investigations were, therefore, undertaken with the following objectives.

1. Critical diesease surveys in various rice growing districts of Manipur should be made. Diseased specimen should be collected.

2. Isolation of fungi associated with collar rot samples collected from various localities and purification and testing their pathogenicity.

3. Effect of different culture media, pH, carbon sources and amino acids on the growth and sporulation of fungi associated with collar rot of rice.

4. Effect of light, temperature and R.H. on the germination of fungi causing collar rot.

5. Effect of infection singly or in combination of fungi associated with collar rot on the severity of the disease.

6. Biochemical changes in the collar infected with collar rot fungi (Total soluble sugars, reducing sugars, non-reducing sugars and free amino acids).

7. Screening of the available high yielding and local rice cultivars of Manipur against collar rot.

8. A comparative efficacy trial on fungicides against the collar rot in *in vitro* and *in vivo* conditions.

Chapter 2

Review of Literature

Studies on rice collar rot disease is reviewed under the following heads :-

1. Distribution of collar rot disease of rice.
2. Symptoms of collar rot disease.
3. Fungal Physiology.
4. Inoculation.
5. Biochemical changes induced in plants infected by plant pathogens.
6. Screening of varietal resistance against rice disease.
7. Control.
8. Assessment of loss.

1. DISTRIBUTION OF COLLAR ROT DISEAS E OF RICE :

The occurrence of rice collar rot caused by *Ascochyta* species was first reported from Japan by Hara (1959). It was then reported from Thailand (Kanjanasoon, 1962); Malays (Chin, 1974); South East Asia (Punithalingam, 1980) and Manipur (Singh and Devi, 1991). Further, the occurrence of collar rot of rice caused by *Pestalotiopsis versicolor* Speg. was first reported from Manipur by Singh *et al.* (1995).

2. SYMPTOMS OF COLLAR ROT DISEASE :

The disease first appeared as small brown lesions at the collar joining the leaf blade and leaf sheath during the booting stage and flowering stage of rice. The lesions increase in size and turn dark brown to black resulting in complete rotting of collar tissues. Finally the disease causes the leaves to separate from the sheath and drop off under favourable conditions. Several leaves on the same plant may dry up and fall in this way (Hara, 1959; Kanjanasoon, 1962; Chin, 1974; Singh and Devi, 1991). In advance stage, the lesions lengthen towards leaf sheath and often rot 1/3 portion of the leaf sheath (Singh and Devi, 1991).

3. FUNGAL PHYSIOLOGY :

a). Effect of Media on the Growth and Sporulation :

Kurosawa (1929) reported that Richards and Knop's solutions were common nutrient media for *Fusarium moniliforme*. Yu and Sun (1978) studied the influence of nutrition on sporulation of *F. moniliforme* and found that rice straw, soyabean meal agar and Sach's agar plus rice straw were best for microcondria production and glucose casamino acid agar and rice straw casamino acid agar for macrocondria production.

Hashioka and Makino (1956) observed the poor growth of *Monographella albescens* on synthetic media and luxuriant growth on potato agar and a special medium containing vitamin B.

Tasugi and Ikeda (1956) reported that potato dextrose agar and oat meal agar were the best nutrient media for promoting good growth and sporulation of *Acrocylindrium oryzae*. Further, they observed good sporualtion in Czapek broth medium.

Swamy and Mani (1978) studied sporulation of *Pestalotia palmarum* Cooke in culture and found that best fruting was obtained on Czapek's agar with yeast extract and under black light.

Chin and Thseng (1981) found that maize agar and straw agar were the best media for good growth of *Sarocladium oryzae*. It was also observed that the best medium for growth of the fungus was potato dextrose agar and the extracts of flag leaf sheath and young panicle stimulated spore germination (Zhuge and Lapis, 1985).

Devi, (1991) studies the effect of different culture media on growth and sporulation of three species of *Fusarium* causing sheath rot of rice and found that mycelial growth of *F. avenaceum*, *F. moniliforme* and *F. graminearum* wa s maximum in potato dextrose broth followed by host extract and Richard's but least in Sabouraud's medium. Further, it was observed that those of synthetic media supported poor growth.

b). Effect of pH on the Growth and Sporulation :

Ramakrishnan (1971) reported that the best pH for good mycelial growth of *Sarocladium oryzae* was at 6.4. It was also observed a wide pH range of 5-8 with optimum 5.5-6.0 for meycelial growth of the fungus (Datta and Purkayastha, 1978). Similarly, Zhuge and Lapis (1985) observed a wide pH range at 3-9 with an optimum at 5.5 for good growth and sporulation.

Gupta *et al.* (1^72) recorded best growth and sclerotial formation of *Macrophomina phaseolina* at pH 6.0.

Venkatasub baiah (1985) reported that in sterilized and unsterilized soils radial growth of the pathogen (*Rhizoctonia solani*) increased with increasing pH to a maximum at pH 7, declining at pH 8, and in pot tests with coffee seedlings damping off was similarly affected by pH levels.

Devi (1991) reported that optimum pH for maximum growth of *Fusarium avenaceum*, *F. moniliforme* and *F. graminearum* was 12.0, 7.0 and 12.0 respectively. Maximum sporulation of *F. avenaceum* , *F. moniliforme* and *F. graminearum* was found at pH 6.5, 11.0 and 7.5 respectively.

Nongdren Khomba (1993) obtained a good fungal growth of

Pestalotiopsis versicolor in a pH range of 5.0-6.5 but maximum at pH5.0. No sporulation occured in all pH upto 7 days of incubation.

c). Effect of Carbon and Nitrogen Sources on Growth and Sporulation :

Thomas (1937) reported that ammonium sulphate or asparagin stiumulated the growth of *F. moniliforme* in culture. Yu and Sun (1978) observed that Xylose was the most suitable carbohydrate for the macro-conidia production of *F. moniliforme*.

Wooster and Cheldelin (1945) found that potassium nitrate was a poor source for the growth of *Penicillium*.

Lilly and Barnett (1951) reported that L-sorbose has been found to be a good to excellent source for the sporulation of *Neocosmospora vasinfecta*. And also found glycine to be a good for a number of fungi. L-arginine and L-asparagine were reported to be favourable sources for sporulation of a large number of imperfect fungi.

Patel *et al.* (1951) reported ammoium sulphate to be a poor source of sporulation of *Pestalotia psidii*.

Tandon and Bilgrami (1957) reported that D-glucose and D-fructose were found to be efficient sources for sporulation of several species of *Pestalotia* and *Pestalotiopsis*. D-mannose has been found to be a moderate to good source for *P. mangiferae*. Dl-*valine* was good source for the sporulation of *P. mangiferae* but this fungus were incapable of reproduction on glycine.

Thind and Randhawa (1957) found that ammonium sulphate to be a poor source for sporulation of some species of *Colletotrichum*. And also have reported some species of *Pestalotiopsis, Colletotrichum* and *Gloeosporium* to sporulate poorly on starch.

Agarwal and Ganguli (1960) have listed that D-glucose and D-fructose are to be efficient sources for sporulation of several species of *Pestalotia* and *Pestalotiopsis*. Maltose and sucrose induced moderate to good sporulation of *Pestalotiopsis versicolor*. L-cystine supported poor sporulation of this fungus and also starch gave poor result for *Pestalotiopsis*. Similar observation has been made by Agarwala (1955) for *Pestalotiopsis, Colletotrichum* and *Gloeosporium* respectively.

Chandra (1961) reported that some species of *Pestalotiopsis, Colletotrichum* and *Gloeosporium* to sporulate poorly on starch.

Bhargava (1962) reported that *Pestalotia glandicola* sporulate profusely on Raffinose.

Prasad (1963) found that L-sorbose supported to be a good to excellent source for the sporulation of *Pestalotia pauciseta*. He studied D-glucose and D-fructose to be efficient sources for sporulation of several species of *Pestalotia*. Maltose and sucrose induced moderate to good sporulation of *P. Pauciseta*. Starch has a good source for sporulation of *P. pauciseta*.

Tandon and Mitra (1963) found that D-fructose and D-glucose to be efficient sources for sporulation of *Pestalotia* sp. and *Pestalotiopsis* sp.

Lal and Tandon (1964) reported that L-sorbose has been found to be a good to excellent source for the sporulation of *Pestalotia bicolor* but raffinose induced poor sporulation of this species.

Roy (1965) reported that maltose and sucrose induce moderate to good sporulation of *Pestalotia albo-maculans*. L-arginine and L-asparagine have also been reported to be favourable sources for sporulation of a large number of imperfect fungi.

Srivastava (1965) reported that *Pestalotia algeriensis* sporulate poorly on arabinose. This fungus sporulate profusely on raffinose.

Starch also has been found to be a good to excellent source for sporulation of *Pestalotia pauciseta* and *P. glandicola*.

Dube (1966) recorded that arabinose was moderate to good source for sporulation of *Pestalotiopsis gracilis*. *Pestalotia* and *Pestalotiopsis* sporulated satisfactorily on D-glucose and D-fructose respectively.

D-mannose was poor to good source for sporulation of *Pestalotiopsis paeoniae*, *P. royenae* and *P. gracilis* respectively. L-sorbose was usually found to be a good to excellent source for the sporulation of *Pestalotiopsis royenae*. Maltose and sucrose induced moderate to good sporulation of *P. royenae*. *P paeoniae* sporulate profusely on raffinose. Starch was excellent to induce sporualtion of *P. pauciseta*, *P. glandicola* and *Pestalotiopsis royenae*.

Purohit (1972) reported that L-arabinose, D-glucose, D-fructose, maltose, sucrose, raffinose, starch, DL-valine, glycine, L-arginine, urea, and L-asparagine were moderate to good sources of carbon and nitrogen for the sporulation of different species of *Pestalotia* under study. However, spore production of *P. macrospora* and *P. aeruginea* was poor on starch and urea respectively. Similar behaviour of *Pestalotia* sp. was observed on these two substances. The pattern of sporulation on D-mannose and L-sorbose was not uniform. D-mannose was poor for *P. versicolor III*; good for *P.* sp., *P. adjusta*, *P. aeruginea*, *P. carveri*, *P. macrospora*, *P. olivacea*, *P. osyridis*, *P. gracilis*, *P. versicolor I*, *P. versicolor II* and excellent for *P. theae*. L-sorbose was excellent for *Pestalotia osyridis*; good for *P. adusta*, *P. carveri*, *P. olivacea*, *P. gracilis*, *P. versicolor* I, *P. versicolor* II; moderate for *P. aeruginea*, *P. theae* and *P. versicolor* III. *Pestalotia* sp. and *P. macrospora*, however, failed to sporulate on this source. The sporulation was, however, initiated and it gradually increased when it was supplied in association with D-glucose in different ratios (3 :1, 2: 2 and 1:3). L-cystine and ammonium sulphate were poor nitrogen sources for most of the species of *Pestalotia* under study.

Das *et al.* (1985) observed sucrose as carbohydrate which significantly promoted mycelial growth of *Pestalotia palmarum* followed by mannitol, glucose, lactose, fructose and starch while sporulation was maximum in lactose and glucose media.

d). Effect of Light on the Growth and Sporulation:

Swamy and Mani, (1978) reported that the fungus *Pestalotia palmarum* Cooke required light for formation of acervuli in culture. Best fruiting was obtained on Czapek's agar with yeast extract and under black light. Fruiting could also be induced by daylight fluorescent and similar under blue, green, yellow and red filters.

Sun and Snyder, (1978) found a good sporulation of *F. moniliforms* when the culture was kept under continuous light. However, large quantities of macroconidia producation were favoured when the culture was exposed to darkness (Yu and Sun, 1978).

Zhuge and Lapie (1985) observed that continuous light had inhibitory effect on growth and sporulation of *S. oryzae* but continuous darkness favoured good growth and sporulation.

Devi and Singh (1994) reported that *Fusarium avenaceum* and *F. graminearum* required complete darkness for their maximum mycelial growth and sporulation. However, *F. moniliforme* required continuous light for its maximum growth. Alternate light and darkness showed synergistic effect on spore production of *F moniliforme*.

e). Effect of Temperature on Growth and Sporulation:

Kurosawa (1929) studied the effect of temperature on growth of *F. moniliforme* and found that the optimum temperature for mycelial growth of the fungus was about 27-30°C, maximum, 36-40°C and minimum, 7-8°C respectively.

Kawamura (1940) reported that the best growth of *A. oryzae* was at 30-310°C and little growth at 37°C but not at all at 13° C and 50°C. It was also reported that good mycelial growth of the fungus at 20-280°C (Ramakrishnan, 1971) and at 24-32°C (Shahjahan *et al.*,1977). However, Zhuge and Lapis (1985) observed a wide temperature range of 15-35°C for growth of the fungus with optimum of 30°C.

Hashioka and Makino (1956) observed a wide temperature range of 20-27°C for growth of *M. albescens* with maximum growth at 20°C. However, several workers reported the optimum temperature for mycelial growth of the fungus as 24°C (Kwon *et al.*, 1973); 24-27°C (Naito and Koshimizu, 1974); 25-30°C (Sittichai, 1974) and 28°C (Ou *et al.*, 1978).

Devi and Singh (1994) reported that the growth of sheath rot fungi was favoured by a wide temperature range. *F. avenaceum* showed an optimum temperature of 25°C for maximum growth. At 30°C, *F. moniliforme* and *F. graminearum* produced maximum mycelial growth.

f). Effect of Relative Humidity (R.H.) on the Growth and Sporulation:

Devi (1991) observed that *F. avenaceum* and *F. moniliforme* started germination at relative humidity of 40 and 60 per cent after 24 hours of incubation and *F. graminearum* at relative humidity of 50 per cent after 40 hours of incubation. Maximum spore germination of *F. avenaceum*, *F. moniliforme* and *F. graminearum* was recorded to be 96, 93 and 92 per cent respectively at 75, 100 and 75 per cent relative humidity.

4. INOCULATION:

Sherf *et al.* (1947) found that mycelial fragments could be used as inoculum, though they are not as infectious as are conidia. The inoculum was prepared by growing the fungus in liquid media. The mycelium was collected by filtering, drying at 40°C for 24 hours and powdering into dust form. The mycelial dust was mixed with limestone (500 mesh) just before inoculation.

Kanjanasoon (1965) observed that 72 hours after inoculation of rice seeds with a spore mass of *Bakanae*, the mycelial of the fungus grew into dense mass between the crown and leaf sheath surrounding the crown.

Bakr and Miah (1975) inoculated detached leaves in a tray by pin-pricking the midrib and putting small blocks of a mycelial culture of *Monographella albescens* on the wounds and keeping them moist. Similarly, Yamaguchi and Ito (1975) reported that spray inoculation with a *conidial* suspension was not successful but infection could be accomplished by adding 1% polypeptone and incubating for 36 hours.

Mukherjee *et al*., (1981) attempted three inoculation techniques on rice variety Pusa 2-21 and found that single grain culture inoculation was most effective for sheath rot (*Acrocylindrium oryzae*).

Devi (1991) attempted five different inoculation methods on KD, 2-6-3, capillary tube inoculation was found to be best in producing highest percentage of infection with typical sheath rot symptoms followed by grain and agar disc inoculation.

a). Effect of Infection or in Combination of Fungi Associated with Rice Disease on the Severity of the Disease:

Devi (1991) observed that either *Fusarium avenaceum* or *F. moniliforme* or *F. graminearum* when inoculated artificially on KD, 2-6-3 rice plants produced typical sheath rot symptoms. The length of diseased lesions was more in rice plants which were inoculated singly with each of the fungus than those on inoculated plants with different combinations. However, combination of *F. avenaceum* + *F. moniliforme* (*F. m.*); *F. m.* + *F. graminearum* (*F. g.*); *F. g.* + *F.*

avenaceum and *F. avenaceum* + *F. g.* + *F. m.* showed antagonistic effect as shown by the production of minimum length of diseased lesion as compared with those produced by individual fungus.

5. BIOCHEMICAL CHANGES INDUCED IN PLANTS INFECTED BY PLANT PATHOGEN:

Iboton *et al.*, (1982) found an increase in total, reducing and non-reducing sugars, starch and protein in greening-infected grapefruit seedlings.

Yadav *et.al.* (1986) reported that out of the fifteen amino acids detected in the experiment only three amino acids namely; lysine, valine and alanine exhibited their presence in both healthy and diseased leaves of sesame infected by *Corynespora cesiecula*. In addition to this, healthy leaves indicated presence of 6 more amino acids, viz. β-amino-n-butyric acid, tryptophan, α-amino-β-butyric acid, proline, arginine and asparagine while the diseased leaves showed the presence of another six amino acids, viz. serine, threonine, tyrosine, histidine, β-alanine, and unidentified amino acid.

Naik *et at.* (1988) observed that there was a marked decrease in total chlorophyll, carotinoids, sugars, phenols and ascorbic acid in betelvine varieties infected with *Colletotrichum gloeosporioides* in the necrotic areas of both the varieties.

Pundir *et al.* (1990) reported that the concentration of total free amino acids increased considerably in infected leaves as compared to healthy leaves. However, no quantitative change in free amino acids was found in the infected leaves.

Prasad *et al.* (1990) found that total sugar, non-reducing sugars and total free amino acids tended to decrease in the seeds with increase in R.H. level and prolongation of the storage period. Reducing sugars increase in similar way.

6. SCREENING OF VARIETAL RESISTANCE AGAINST DISEASE:

Subramanian and Ramakrishnan (1975) studied the incidence of sheath rot on different rice cultivers under field conditions and recorded Annapoorna was comparatively more susceptible to the disease than TN-1, TK-6, Tella himsa, Kanto, and Segadis. Amin (1976) found that twelve tall rice varieties and twenty advanced dwarf lines of 1Q,1,544 were highly resistant to sheath rot fungus, *Sarocladium oryzae* under natural conditions.

Mukerjee *et al.* (1981) screened 169 indigenous germ plasms of rice by inoculating with single grain culture of *Acrocylendrium oryzae* but none was found immune to sheath rot disease.

Chien and Thseng (1982) studied the susceptibility of rice varieties to sheath rot fungus both in field and glasshouse and found that *indica* varieties were more susceptible than *japonica* but lowest in Taichung 61.

Lewin and Vidhyasekaran (1987) observed that 10 rice varieties out of 59 were resistant against the sheath rot in the field under very high disease pressure. In a field screening study, 8 out of 34 rice cultivars and lines immune to *Fusarium* sheath rot disease. However, none of the 8 lines was immune in pot experiment but had lower disease scores (Grewal and Kang, 1988).

Iqbal *et al.* (1990) conducted a study on *Ascochyta* blight reactions of lentil genotypes under artificial epiphytotic conditions in the field to identify sources of resistance to lentil blight and their possible use in breeding for disease resistance. The disease severity was recorded out of 9 scale. Out of 152 genotypes tested 17 cultivars were highly resistant, 40 were resistant, and 34 showed an average reaction while the rest were susceptible to highly susceptible.

Kapoor *et al.* (1990) evaluated lentil lines resistant to *Ascochyta* blight. The variaties HPL5 and LG 171 expressed a disease score of 1 and were highly resistant to *Ascochyta* blight. The 13 LG advanced breeding lines (169, 174, 191, 195, 201, 209, 217, 218, 219, 220, 221 and 225) showed a disease score of 2 or below 2 on a 1-9 point rating scale and were resistant. Nine lines showed a disease score at 2. 1-2. 87. Out of 1,047 germ plasm line tested, L 442 and L 448 consistantly showed a disease score of 1 for two years and were highly resistant. The seven lines (293, 434, 462, 511, 512, 521 and 624) gave a disease score of 1.33-1.7 and were highly resistant to resistant. Another eight lines expressed a disease score of 2-2.6. The seven LG lines (169, 170, 172, 173, 191 and 209) gave stable resistance.

Devi (1991) found that twentyfive valley rice varieties of Manipur were scored as 3 highly resistant, 13 resistant, and 6 highly susceptible to sheath rot. Among hill varieties, 12 showed highly resistant reaction, 14 resistant, 9 susceptible and 7 highly susceptible. Further, among breeding lines obtained from Rice Directorate, Hyderabad, 8 expressed highly resistant, 7 resistant, 6 susceptible and 4 highly susceptible reaction respectively to sheath rot.

Gill *et al.* (1993) observed that among the fine grain rice varieties/lines tested against bakanae and foot rot disease, Basmati, 370, 4439, 1053-1-2, 1053-32 and 33608 showed resistance under low inoculum potential. These varieties can be grown in fields with low soil infestation. Basmati 385, Basmati 6129, and 4048 are susceptible under all conditions and should be avoided. Medium grain varieties/ lines IR6, IR8, IR9 and Shadab, KS282, PK3303-15-1, and PK1399-12-1-1-0-6 were highly resistant and can be grown successfully in severely infested fields.

7. CONTROL:

Chinnaswamy *et al.* (1981) observed the effective control of sheath rot of rice by treating with certain fungicides, all 6 treatments significantly reduced the incidence of *Sarocladium oryzae*. Bavistin (carbendazim) was the most effective, followed by HMP, MBC (carbendazim), Aureofungin and Difolatan (captafol).

Reddy *et al.* (1984) found that *Acrocylindrium oryzae* was significantly inhibited by Bavistin followed by Blue copper and Antracol in *in vitro.*

Upadhyay (1984) reported that phygon XL (dichlone) proved most effective against *Pestalotiopsis funerea in vitro* causing leaf spot of *Ecalyptus globulus* followed by Difolatan 80W (captafol), Dithane Z-78 (Zineb), Zincop and Cupramar (Copper oxychloride). All except Difolatan were less effective after successive prolonged incubation. At lower concentration of the antibiotics tested, Mycostatin was most effective followed by Streptopenicillin, Griseofulvin, Streptomycin Sulphate and Benzyl Penicillin.

Das and Mahanta (1985) reported that Bavistin (carbendazim). Tecto-60 (thiabendazole) and Hexathir (thiram) completely inhibited *P.* (*Pestalotiopsis*) *palmarum* in *in vitro*.

Gaerber and Motte (1990) tested 14 fungicides against *Ascochyta fabae* Speg. and Befran was found most effective. The effectiveness reached 78 to 96 per cent even at a very low concentration (0.0007 per cent).

Devi (1991) found that yield significantly increased due to the application of most fungicides against fusarial sheath rot of rice. Maximum yield were recorded with Dithane, M-45, closely followed by Bavistin and Benlate. The lowest rice yields were found in Kitazin sprayed plots.

8. ASSESSMENT OF LOSS:

Prabakaran *et al.* (1974) recorded 85% yield loss due to *Acrocylindrium oryzae* infection in Tamilnadu whereas 57% damage was reported by Mohan and Subramanian (1977).

Chakraverty and Biswas (1978) observed that the yield loss due to sheath rot on IR-2588-19-1-22 and B54-Kn-58-5-3 were 10% and 26% respectively under natural conditions. A potent reduction of 52.8% in the total yield of rice was recorded due to infection by *S. oryzae* on IR 442-58 in irrigated plots (Estrada *et al.*, 1984).

Tikoo (1985) reported that due to severe incidence of sheath rot on CV-K-78-13, the number of spikeletes/panicle; number of grains/panicle and 1,000 grain weight were reduced markedly resulting in loss of crop yield.

Sugha and Singh (1989) studied the effect of glume blight of rice on grain yield and quality and found that progressive increase in disease. Severity resulted in a corresponding decrease in the weight of whole and milled grains, increase in length-width ratio of grains.

Dhindra *et al*. (1991) found that maximum percentage of false smutted tillers and grains increase in chaffiness and decrease in grain weight were recorded in rice CV. PR 109 and least in PR 103.

Chapter 3

Materials and Methods

The present investigations were undertaken in the Department of Plant Pathology, College of Agriculture, Central Agricultural University, Iroisemba, Imphal Manipur from 1993-1995. The experimental approaches and procedures adopted during the course of investigations are given below.

3.1 COLLECTION OF DISEASE SPECIMEN:

During October - November, 1993 and 1994, critical disease surveys were conducted at several rice growing areas in Imphal, Chandel, Senapati, Thoubal and Bishnupur districts of Manipur. Young or old rice leaves showing typical collar rot symptoms were collected from these districts and brought to the Laboratory of The Plant Pathology for isolation.

3.2 ISOLATION OF FUNGI ASSOCIATED WITH RICE COLLAR ROT:

The collected rice collar rot samples were cut into small bits. The bits were surface sterilized with 1 per cent sodium hypochlorite solution for 2 - 3 minutes, rinsed in sterile distilled water once and plated on Potato Dextrose Agar (PDA). The inoculated dishes were incubated at 25 + 1^{0}C. The fungal cultures were purified by hyphal tip culture and maintained on Potato Dextrose Agar (PDA) slants by subculturing time to time.

3.2.a. Pathogenicity Tests:

The pathogenicity tests of collar rot fungi were carried out by the method of Devi (1991)

3.2.b Preparation of Mycelial Suspension :

Mycelial mats of 7-day-old culture of *Pestalotiopsis versicolor* (Speg.) and *Chaetomium globosum* growing in petri dishes (11.0.cm) were collected separately and transferred to an electric blender containing 50 ml of sterilized water, The blender was run in alternate cycles of high and low speed for 4 to 5 minutes. The mycelial suspensions thus prepared were centrifuged at 10,000 rpm and then at 15,000 rpm for 10 minutes each under aseptic conditions. Each pellet was resuspended in sterile distilled water and diluted serially to give $1\text{-}1.5 \times 10^3$ infective propagules of 170-250 µm size per ml. The slicing and beating action of the blades followed by differential centrifugations reduced the coarse fungal mats into small enough to transverse through a fine spray nozzle of an atomiser. The mycelial suspension were thus prepared for effective inoculation.

3.2.c. Preparation of Spore Suspension :

Huge amount of spores of collar rot fungi (*P. versicolor* and *C. globosum*) were collected separately from 20-day-old culture growing in petri dishes (11.0.cm) and transferred to an electric blender containing 50 ml of sterilized water. The blender was run in alternate cycles of high and low speed for 4 to 5 minutes, which properly homogenized the spore masses. With the help of haemocytometer, the number of spore per ml was calculated.

3.2.d Distribution of Fungi Associated with Collar Rot at various Localities in Different Districts of Manipur

During the two crop seasons (1993 and 1994), several visits were made to Lamsang, Lamphelpat, Heibongpokpi, Sawombung,

Lamlai, Langjing, Khumbong, Keirao, Heingang, Khonghampat and Langthabal of Imphal district; Thoubal Kiyam, Wangbal, Ngangoupat, Waithou, Wangjing, Yairipok, Khongjom, Kakching, Kakching Khunou of Thoubal district; Sugnu, Pallel and Chandel of Chandel district; Nambol Oinam, Bishnupur, Moirang and Wangoo of Bishnupur district and Sekmai of Senapati district. Diseased specimens were collected from these localities and fungi associated with collar rot were isolated for studying the distribution of these fungi.

3.3.1. Effect of Media on the Growth and Sporulation of Fungi Associated with Collar Rot of Rice :

Seven different culture media, namely Asthana & Hawker's medium, Potato dextrose, Oatmeal, Czapek dox medium, Richards's medium, Corn meal and Sabouraud's conservation were prepared as given by Riker and Riker (1936). Fifty ml of medium was dispensed in each Erlenmeyer conical flask (100 ml cap.). A 5 mm mycelial plug taken out from the actively growing colonies of 3-day-old culture was transferred aseptically to each flask. The inoculated flasks were then incubated at 25+1^{0}C for 7 days. Each treatment was replicated for 8 times (5 for growth and other 3 for spore production studies). After incubation period, the mycelial mats were harvested by filtering through pre-weighed No. 1 (11.0.cm diameter) and dried at 60^{0}C for 72 hours in a hot air oven and then re-weighed to estimate fungal growth on dry weight basis.

From the remaining three flasks, the extend of spore production of *P. versicolor* and *C. globosum* was studied after 7 days of incubation after properly homogenized the fungal masses using electric blender. The number of spores per ml was calculated with the help of haemocytometer.

3.3.2. Effect of pH on the Growth and Sporulation of Fungi Associated with Collar Rot of Rice :

Studies were conducted on Richard's broth medium. Eleven

pH values, namely 2.0, 3.0, 4.0, 5.0, 6.0, 7.0, 8.0, 9.0, 10.0., 11.0 and 12.0 were adjusted with pH meter (Systronics) using either KOH or HCl 50 ml of Richard's broth was dispensed in each 100 ml Erlenmeyer conical flask and autoclaved at 121°C for 20 minutes. All the treatments were replicated 8 times. Then flasks were inoculated with a 3-day-old mycelial plug of 5 mm size and incubated at 25+1°C for 7 days. Mycelial mats were harvested on pre-weighed Whatman No.1 filter paper and dried at 60°C for 72 hours using hot air oven and re-weighed. Fungal growth was determined on dry weight basis. The number of spores contained in 1 ml of suspension was also calculated with the help of haemocytometer.

3.3.3. Effect of Different Carbon Sources and Amino Acids on the Growth and Sporulation of Fungi Associated with Rice Collar Rot :

Ten carbon sources namely D(-) Arabinose; Dextrose; D(-) Fructose; Mannitol; D(+) Xylose; D(+) Maltose; Sucrose; D(-) Ribose; D(+) Mannose; Cellobiose and twentyone amino acids namely Glycine; DL-Methionine; L-Glutamic acid; DL-Alanine; L-Cysteine; DL-Valine; L-Lucine; DL-nor-Leucine; DL-Aspartic acid; DL-Ornithine; L-Histionine; DL-Threonine; DL-Iso-Leucine; L-Hydroxyproline; L-Arginine; DL-B-Phenylalanine; DL-Tryptophan; DL-Serine; L-Proline; DL-2-Aminobutyric acid and L-Leucine were used for study. Asthana and Hawker's medium was used as basal medium. Glucose and Potassium nitrate content in the basal medium were substituted by different carbon sources and amino acids mentioned above by adding separately to 50ml basal medium to give an equivalent amount of carbon (2 g C/1) and nitrogen (0.485 g N/1). All treatments were replicated 8 times (5 for growth and other 3 for sporulation). The media without carbon and nitrogen sources served as control. Fungal growth was determined on dry weight basis after 7 days of incubation. The number of spores/ml was calculated with the help of haemocytometer.

3.3.4. Effect of Light on Growth and Sporulation of Fungi Associated with Collar Rot of Rice :

From actively growing culture of collar rot fungi, 5 mm fungal plugs were cut aseptically from actively growing cultures and transferred to each 100 ml Erlenmeyer conical flask containing 50 ml potato dextrose broth. The inoculated conical flask were exposed to three different conditions of light namely; complete light, alternate light and darkness (12h/12h) and complete darkness for 7 days at 25+1°C. All the treatment were replicated 8 times (5 replications for growth and 3 for sporulation). The growth and spore production of collar rot fungi were determined as described in 3.3.1.

3.3.5. Effect of Temperature on the Growth and Sporulation of Fungi causing Collar Rot :

From pure cultures of collar rot fungi, 5 mm mycelial plugs were cut aseptically from the actively growing cultures and transferred to each 100ml. Erlenmeyer conical flask containing 50 ml of each potato dextrose broth. The inoculated conical flasks were exposed to different temperatures such as 5, 10. 15, 20, 25, 30, 35 and 38°C for 7 days. All treatments were replicated 8 times. The growth and spore production of collar rot fungi were determined similarly as described in 3.3.1.

3.3.6 Effect of Relative Humidity (R.H.) on the Germination of Fungi Associated with Rice Collar Rot:

Different relative humidities were prepared with the help of concentrated sulphuric acid and distilled water. The mixing ratios of sulphuric acid with distilled water described by Hosokawa and Kubota (1957) were followed with little modification.

The different mixtures were poured into separate cavity blocks (10 ml cap.). Two drops of spore suspension ($4X10^3$ spores/ml) of different collar rot fungi were put on clean cavity slides. The cavity blocks holding the mixtures were placed inside large petri dishes and the cavity slides were then placed on the cavity blocks. After that, the petri dishes were covered with lids and sealed with veseline to maintain the required relative humidities inside the different petri dishes. The petri dishes were incubated at 25+1°C for 24, 48 and 72 hours to study the effect of different relative humidities on spore germination of fungi associated with rice collar rot.

3.4.1. Effect of Infection Singly or in Combination of Fungi Associated with Collar Rot on the Severity of the Disease :

Enough rice plants (var. Leima Phou) grown in the field were used for the studies. Fungi namely *P. versicolor* and *C. globosum* were subcultured on PDA for inoculation on rice collar region.

Inoculations:

Leima Phou (KD, 2-6-3) rice plants grown in the field/pots were inoculated singly and in combination of fungi by placing the mycelial plug on collar region of rice plant @ 1 fungal plug/collar. The entire inoculated collars were wrapped with a polyethylene tape to keep moist. Each treatment was replicated for 3 times. The length of disease lesions were recorded.

3.4.2. Effect of First Collar, First and Second and First, Second and Third Affected Collars when Inoculated with either *P. Versicolor* or *C. Globosum* on Rice Yield :

Leima Phou (KD, 2-6-3) rice plants grown in field were inoculated with a single fungal plug of either *P. versicolor* or *C. globosum* by placing on first collar, first and second collar, and first, second and third collar region of the plant at rice booting stage. Each of the entire inoculated collar was wrapped with a polyethylene tape. Each treatment was replicated for 3 times. The uninoculated Leima Phou (KD, 2-6-3) rice plant was taken as control. Yield was recorded separately by taking the weight of 1,000 grains (g).

3.4.3. Effect of Artificial Inoculation of Collar Rot Fungi on Rice Yield :

A large number of rice cultivars Taichung, Leima Phou (KD, 2-6-3) and China dwarf grown in the fields/pots were used for this experiment. To study the reaction of rice plants against the fungi, mycelial fragments were prepared as similarly described in 3.2.b. A drop of mycelial suspension was placed on the collar region with the help of sterilized plastic dropper bottle. The inoculated plants were covered with perforated polythene bags to maintain higher humidity inside for 24 hours. Disease data was recorded and rice yield were estimated.

3.5. BIOCHEMICAL CHANGES IN THE COLLAR INFECTED WITH COLLAR ROT FUNGI (TOTAL SOLUBLE SUGARS, REDUCING SUGARS, NON-REDUCING SUGARS AND FREE AMINO ACIDS):

Diseased collars caused by *P. versicolor* and *C. globosum* and healthy collars of rice were collected in separate bags and brought to the laboratory of Plant Pathology. Small bits were cut from collar tissues. The bits were placed in paper bags separately and dried at 65°C in a hot oven to a constant weight. The dried material was powdered and stored in stoppered glass vials for biochemical analysis.

3.5.1 Extraction for Amino Acids, Total Sugars, Reducing and Non-Reducing Sugars :

A gram of the powdered tissue material from collar rot infected and healthy collar was extracted separately with a pestle and mortar in a small volume of 70% ethanol. The extract was filtered through whatman No. 42 filter paper. This process was repeated thrice for complete extraction of the alcohol soluble components. The combined filtrates were evaporated to dryness on a water bath at 70-75°C. Each residue was redissolved in a known volume of ethanol (6 ml of 20% ethanol). The alcoholic extract was centrifuged at 6000 rpm for 15 minutes. The supernatants thus obtained were designated as extract "A" (disease caused by *P. versicolor*) 'B' (disease caused by *C. globosum*) and 'C' (healthy) and preserved in refrigerator and used for analysis of the amino acids, total sugars, reducing and non-reducing sugars.

3.5.1.a. Estimation of Amino Acids :

Amino acids were determined using two dimensional descending paper chromatographic method of Mahadevan (1974). A volume of 9.5 ml each was taken from extract 'A', 'B' and 'C' was spotted individually on Whatman No. 1 filter papers (28X28 sq. cm) with the help of micropipettes. The spotted papers were run two dimensionally with two solvent systems namely phenol saturated with water (74 g: 25 ml) and n-butanol : glacial acetic acid: water (4: 1:5 V/V/V) for about 10 hours in a paper chromatographic chamber saturated with solvent vapours. After developing, the chromatograms were air dried at room temperature and sprayed with ninhydrin solution. This solution was prepared by dissolving 300 mg of commercial ninhydrin in 100 ml n-butanol. After spraying, the chromatograms were heated at 110°C for 10 minutes in a hot air oven. The ninhydrin positive spots were identified after comparing them against pure amino acids run and developed under identical conditions.

The ninhydrin positive spots were cut into pieces and eluted individually at 6 ml. of 70 per cent ethanol for 3 hours and the absorbance was measured at 570 nm. in a colorimeter using 70 per cent ethanol as a blank. The quantities of individual amino acids were worked out separately against the standard curve prepared with glycine.

3.5.1.b. Estimation - Total Sugar:

The procedure, suggested by Ketiku (1973) was used for the estimation of total sugars. A volume of 0.1 ml. was taken from extract 'A', 'B' and 'C' into a clean test tube and diluted by adding

9.9 ml. distilled water. One ml each of the diluted solution was then taken into a clean test tube in which 1 ml of 5% phenol solution and 5 ml of concentrated sulphuric acid were added. After an hour, the optical density of the yellow brown solution was measured on Bausch and Lamb spectronic-20 at 480 nm. All the reagents minus sample were used as blank. The total quantity of the sugars of each sample was calculated against a standard curve prepared by using glucose (AR).

3.5.1.c. Reducing and Non-Reducing Sugars :

The reducing and non-reducing sugars were estimated by the colorimetric method of Somogyi (1952) using Nelson's chromogenic reagent (Nelson, 1944).

Somogyi's reagent $CuSo_4 5H_2O$	4.0 g
Na_2CO_3 (anhydrous)	6.0 g
Rochelle salt	
(sodium potassium tartarate)	3.0 g
$NaHCO_3$	4.0 g
$NaSO_4$ (anhydrous)	45.0 g
Distilled water to make upto 250 ml.	

Four grams of $CuSO_4$ were dissolved in 25 ml. of distilled water and Na_2CO_3 and Rochelle salt in 50 ml of distilled water; sodium sulphate was dissolved in 130 ml distilled water in a beaker which was just near to boiling. The beaker was put on a hot plate and the solution was kept boiling for 2-3 minutes. It was first cooled in running tap water and then kept at room temperature for further cooling. To Na_2CO_3 Rochelle salt solution, 9 ml of $CuSO_4$ solution was added drop by drop while stirring the solution with glass rod. The solution turned to sky blue colour. Thereafter sodium bicarbonate was added to it. Some crystalline material (cuprous oxide) settled down and the solution was filtered through Whatman No. 42 filter paper. The clear solution was mixed with sodium sulphate solution and the final volume was raised to 250 ml with distilled water.

Nelson's Reagent :

Ammonium molybdate measuring 12.5 g was dissolved in 225 ml of distilled water; to it was poured 10.5 ml of concentrated sulphuric acid drop by drop and to this solution was added 1.5 g of sodium arsenate dissolved in 12.5 ml of water. The final solution was put in a brown bottle and kept at 37°C for 48 hours.

Reducing Sugars :

To 0.5 ml each of extract 'A', 'B' and 'C' taken in a test tube was added 0.5 ml of Somogyi reagent followed by a small quantity of distilled water. The tube was put in boiling water of three extracts for 10 minutes. It was taken out and cooled in a pan of cold water. Thereafter 0.5 ml of Nelson's reagent was added. The contents were mixed thoroughly till effervescence ceased. The final volume was made up with water. The optical density was measured at 490 nm.

Non-Reducing Sugars :

For this purpose, the hydrolysis of the extract was done first. To 0.5 ml each of extract 'A', 'B' and 'C' taken in a clean test tube, one ml of IN H_2SO_4 was added. It was heated to 50-55°C for half an hour in water bath and then cooled in a pan of cold water. One or two drops of indicator, methyl red solution were added. The reddish solution was neutralized with 30% NaOH adding it dropwise. The final volume was made to 5 ml with water. From this solution, one ml was taken in a test tube for the estimation as reducing sugars as described above. By sub's' traction, the amount of inverted sugar was determined. This was multiplied by the factor 0.93 to obtain the amount of non-reducing sugars present.

For reducing sugars, standard curve from glucose was prepared.

3.6. SCREENING OF THE AVAILABLE HIGH YIELDING AND LOCAL RICE CULTIVARS OF MANIPUR AGAINST COLLAR ROT:

Field screening trials were conducted during 1993 and 1994 rice season at the experimental field of College of Agriculture (CAU), Iroishemba to evaluate 100 cultivars/ varieties collected from different areas of Manipur, 29 exotic cultivars procured from Rice Research Station, Wangbal, Directorate of Agriculture, Govt. of Manipur.39 rice lines received from Directorate of Rice Research Rajendranagar, Hyderabad-30. All the 168 entries of rice seeds were raised in nursery beds and transplanted after 30 days at the main field in two rows each of 2 m long each maintaining spacing at 20x30 cm plant to plant and row to row between the entries at boot stage using fungal plug (5 mm) taken out from actively growing of 3-day-old culture of collar rot fungi. At every 10 test entries, susceptible check either China dwarf or Leima Phou (KD, 2-6-3) were planted. Disease data were taken after 10 days of inoculation. The test varieties / lines/ cultivars were rated as Resistant, Tolerant, Susceptible and most Susceptible using Standard Evaluation System for rice (IRRI-1985).

3.7.1. Effect of Different Fungicides on Growth and Sporulation of Rice Collar Rot Fungi :

The efficacy of five fungicides on the *in vitro* growth and sporulation of collar rot fungi were studied using poisoned food technique of Sharvelle (1960). The concentrations of each fungicide were adjusted to 100 per cent active ingredient (a.i.) Fungicides namely, Blitox-50 (copper oxychloride, 50% Cu) @ 0.2, 0.05, 0.025%. Foltaf (Ortho 5865 Folcid) N- (1, 1, 2, 2-tetrachloroethyl) Sulfenylcis-4-cyclohexene 1, 2, dicarboximids, 80% wp) @ 0.1, 0.05, 0.025% ; Dithane M-45 (a coordination product of zinc, ion and manganous ethylene bis dithiocarbamate) @ 0.2, 0.05, 0.025%; Bavistin (Methyl 1-2-ben-zimidazole carbamate) @ 0.1, 0.05, 0.025% and Topsin-M (1,2-bis (3-methoxycarbonyl-2-thioureido) benzene) @ 0.1, 0.05, 0.025% were incorporated with 50 ml potato dextrose broth after autoclaving. A 5 mm mycelial plug taken from 3-day-old actively growing each culture was aseptically transferred to each conical flask. Each treatment was replicated 8 times (5 for growth and 3 for sporulation). The inoculated flasks were incubated at 25+1°C for 7 days. The medium without fungicide served as control. The mycelial dry weight was determined similarly as described in 3.3.1

To confirm the fungicidal and fungistatic effects of each fungicide under study, the mycelial plug showing no growth in the fungicids treated medium was taken out and washed in sterilized distilled water (three changes) and transferred aseptically to Potato Dextrose Agar (PDA) slants. These slants were then incubated at 25+1°C for 3-4 days. Fungal growth was then recorded.

3.7.2 Effect of Fungicides on Disease Intensity and Rice Yield :

Two seedlings of China dwarf rice (most susceptible culture) were transplanted in each 12" diameter earthen pots. Five fungicides mentioned in 3.7.1. were evaluated their efficacy against collar rot disease of rice. Two foliar sprays i.e. one at booting and other, just after flowering stages of rice were given after a day of fungal inoculation. The inoculation method described at 3.4.3. was followed. Unsprayed pots served as control. Disease intensity and rice yield data were recorded.

Chapter 4

Experimental Results

4.2 ISOLATION OF FUNGI ASSOCIATED WITH RICE COLLAR ROT :

Two fungi namely, *Pestalotiopsis versicolor* (Speg.) and *Chaetomium globosum* were consistently found to be associated with rice collar rot disease in Manipur.

4.2.1. Pathogenicity Test :

When Leima Phou (KD, 2-6-3) and China dwarf rice plants which were artificially inoculated with each fungus, *P. versicolor* and *C.*

(Fig. 1)

globosum, produced typical collar rot symptoms after 3 days of inoculation (Fig. 1). On re-isolation from the artificially induced collar rot symptoms, the same fungi were found.

4.2.d. Distribution of Fungi Associated with Collar Rot of Rice at various localities in different districts of Manipur :

Both the fungi were well distributed in 174 fields of various localities visited in Imphal, Thoubal, Chandel, Bishnupur and Senapati districts of Manipur (Table 1). It was observed that the frequency of *P. versicolor* was 70 per cent in Imphal district followed by Thoubal, Bishnupur, Chandel and Senapati, whereas *C. globosum* was found 25 per cent in Imphal followed by Bishnupur, Chandel, Thoubal and Senapati while isolated from the collected disease samples.

4.3.1 Effect of Different Media on Growth and Sporulation of Collar Rot Fungi :

It was evident from the results presented in Table 2 that all the media under study supported mycelial growth of both the fungi. However, maximum mycelial growth of the fungus, *P. versicolor* was found in Richard's medium followed by potato dextrose broth but least in Czapek-dox. The fungal growth was not significantly different in Corn meal and Asthana and Hawker's media. Potato dextrose broth supported maximum mycelial growth of the Fungus, *C. globosum* followed by Corn meal but least in Czapek-dox medium. Further it was observed that the test culture media could not induce sporulation of both fungi upto 7 days of incubation.

4.3.2. Effect of Different pH on Growth and Sporulation of Collar Rot Fungi :

Data presented in Table 3 showed that good mycelial growth of *P. versicolor* was obtained in potato dextrose broth at different pH tested but they failed to grow at pH 12. However, maximum fungal growth of *P. versicolor* was obtained at pH 5 followed by pH 6.

TABLE -1

Distribution of Fungi Associated with Collar Rot of Rice at Various Localities in Different Districts of Manipur

S.No.	District	Locality	No. of Fields visited	Frequency of Fungus *P. versicolor* (%)	Frequency of Fungus *C. globosum* (%)
1.	Imphal	Lamsang, Lamphelpat, Heibongpokpi, Sawombung, Lamlai, Langjing, Khumbong, Keirao, Heingang, Khonghampat and Langthabal.	66	70	25
2.	Thoubal	Thoubal Kiyam, Wangbal, Ngangoupat Waithou, Wangjing, Yairipok, Khongjom, Kakching and Kakching Khunou.	54	69	12
3.	Chandel	Sugunu, Pallel and Chandel	18	30	15
4.	Bishnupur	Nambol, Oinam, Bishnupur, Moirang and Wangoo.	30	50	20
5.	Senapati	Sekmai	6	20	10
		Total	174		

TABLE - 2

Effect of Different Media on Growth of Fungi Associated with Rice Collar Rot

Sl.No.	Media	*P. versicolor* (Speg.) growth* (mg)	*C. globosum* growth* (mg)
1.	Asthana & Hawker's Medium	231.60	166.63
2.	Potato dextrose broth	368.69	263.00
3.	Oat meal	52.80	47.88
4.	Czapek dox medium	32.38	33.63
5.	Richard's medium	675.00	82.50
6.	Corn meal	317.50	258.00
7.	Sobouraud's Conservation	45.50	51.88
8.	Control	0.00	0.00
	* Mean of 5 replications CD 5%	114.42	38.99

Similarly mycelium of *C. globosum* grew well at wide pH range (Table 3). The maximum mycelial growth was obtained at pH 8 followed by pH 7 but least at pH 10. However, this fungus failed to grow at pH 12.

TABLE - 3

Effect of Different pH on Growth of Fungi Associated with Rice Collar Rot

Sl.No.	pH	*P. versicolor* (Speg.) growth (mg*)	*C. globosum* growth (mg*)
1.	2.0	250.83	44.67
2.	3.0	363.50	64.83
3.	4.0	405.83	73.50
4.	5.0	448.33	112.67
5.	6.0	412.33	113.13
6.	7.0	391.17	125.53
7.	8.0	384.00	152.30
8.	9.0	383.00	62.33
9.	10.0	376.00	43.67
10.	11.0	318.00	42.33
11.	12.0	0.00	0.00
	CD 5%	36.27	38.22

* Mean of five replication

The mycelial growth of both fungi failed to sporulate upto 7 days of incubation.

4.3.3.a. Effect of Different Carbon Sources on Growth and Sporulation of Collar Rot Fungi :

Results presented in Table 4 revealed that maximum mycelial

growth of *P. versicolor* was found in cellobiose but least growth in Ribose. However, no significant difference was observed in Mannose, Cellobiose, Maltose, Fructose and Mannitol respectively. Further, maximum mycelial growth of *C. globosum* was found in Mannose followed by Sucrose but least growth in Arabinose. However, mycelial growth was not significant different in all the rest carbon sources.

TABLE - 4

Effect of Different Carbon Sources on Growth of Fungi Associated with Rice Collar Rot

Sl.No.	Carbon Sources	*P. versicolor* growth (mg*)	*C. globosum* growth (mg*)
1.	Arabinose	52.50	16.82
2.	Mannitol	151.50	27.67
3.	Fructose	153.50	27.50
4.	Xylose	59.50	35.00
5.	Maltose	149.50	25.82
6.	Ribose	20.00	27.50
7.	Sucrose	129.00	33.00
8.	Mannose	154.00	43.68
9.	Cellobiose	239.00	29.75
10.	Dextrose	39.27	27.00
11.	Control	5.00	5.00

* Mean of five replications

	CD 5%	92.36	11.16

It was also observed that carbon sources under test failed to induce sporulation of the fungi upto 7 days of incubation.

4.3.3.b. Effect of Different Amino Acids on Growth and Sporulation of Collar Rot Fungi :

Results presented in Table 5 showed that good growth (94.63 mg) of *P. versicolor* was found in L-Lucine followed by DL-Aspertic acid (91.88 mg) but poor (15.23 mg) in L-Histidine. In case of *C. globosum*, maximum mycelial growth (39.00 mg) was supported by L-Leucine followed by L-Lucine (27.25 mg) but least (3.00 mg) in DL-Methionine. Amino acids under study could not induce sporulation of both fungi upto 7 days of incubation.

TABLE - 5

Effect of Different Amino Acids on Growth of Fungi Associated with Rice Collar Rot

Sl.No.	Amino Acid	*P. versicolor* growth (mg*)	*C. globosum* growth (mg*)
1.	DL - Methionine	30.08	3.00
2.	L - Glutamic	57.3	20.50
3.	DL - Alanine	87.18	12.33
4.	L - Cysteine	37.00	0.00
5.	DL - Valine	49.88	10.25
6.	L - Lucine	94.63	27.25
7.	DL - Nor - Leucine	28.15	14.50
8.	DL - Aspertic	91.88	22.50
9.	DL - Ornithine	84.20	11.60
10.	L - Histidine	15.23	24.00
11.	DL - Threonine	16.50	16.00
12.	DL - Iso - Leucine	22.75	14.25
13.	L - Hydroxy Proline	17.50	8.15
14.	Glycine	42.75	20.65
15.	L - Arginine	48.25	24.00
16.	DL-β-Phenylalanine	60.50	14.00
17.	DL - Tryptophan	21.50	14.5
18.	DL - Serine	64.25	11.00

19.	L-Proline	60.00	15.75
20.	DL-2-Amino-Butyric	25.50	18.25
21.	L - Leucine	38.00	39.00
22.	Control	8.00	8.00

* Mean of five replications

	CD 5%	11.84	9.09

4.3.4 .Effect of Light on Growth and Sportulation of Collar Rot Fungi :

Different light conditions had effect on growth and sporulation of fungi. *P. versicolor* and *C. globosum* when exposed to complete darkness produced maximum mycelial growth (Table 6). However, *P. versicolor* when exposed to complete light could sporulate after 4 days of incubation. But *C. globosum* when exposed to complete light for 7 days failed to sporulate. It was observed that complete darkness and alternate light and darkness (12h/12h) could not induce sporulation of both fungi upto 7 days of incubation.

TABLE - 6

Effect of Light on Growth and Sporulation of Fungi Associated with Rice Collar Rot

Sl.No.	Condition of light	*P. versicolor* growth (mg*)	*C. Globosum* growth (mg*)	No.of spore/ ml	
				P. versicolor	*C. globosum*
1.	Complete light	45.50	32.46	$3.5x10^4$	0
2.	Alternate light darknes (12h/12h)	57.00	40.65	0	0
3.	Complete darkness	70.00	49.94	0	0

4.3.5. Effect of Temperature on Growth and Sporulation of Fungi Associated with Collar Rot of Rice :

Data presented in Table-7 showed that temperature played and important role in the growth of collar rot fungi. The growth of collar rot fungi was favoured by wide temperature range of 10-35°C. *P. versicolor* and *C. globosum* showed an optimum temperature of 25°C for maximum growth. But both fungi failed to grow at 5° and 38°C.

TABLE - 7

Effect of Temperature on Growth and Sporulation of Fungi Associated with Rice Collar Rot

Sl.No.	Temperature (°C)	Growth (mg)	
		P. versicolor	*C. globosum*
1.	5	-	-
2.	10	111.00	68.24
3.	15	122.01	73.00
4.	20	381.00	114.03
5.	25	446.00	120.90
6.	30	403.20	102.00
7.	35	101.00	52.00
8.	38	-	-

4.3.6 Effect of Relative Humidity (R.H.) on the Germination of Fungi Associated with Rice Collar Rot :

It was observed that spores of *P. versicolor* and *C. globosum* started germination from 50 per cent relative humidity onwards (Table 8). Spores of *P. versicolor* germinated after 35 hours of incubation and those of *C. globosum* after 24 hours of incubation. Cent per cent spore germination of both fungi were recorded at 100 per cent relative humidity.

TABLE - 8

Effect of Relative Humidity on Spore Germination of Fungi Associated with Rice Collar Rot

Sl.No.	Relative Humidity (%)	Spore germination (%)		
		P. versicolor		*C. globosum*
		24h	35h	24h
1.	25	0	0	0
2.	30	0	0	0
3.	35	0	0	0
4.	50	0	2.0	10.0
5.	75	0	56.0	70.0
6.	90	0	80.0	85.0
7.	100	0	100.0	100.00

4.4.1 Effect of Infection Singly or in Combination of Fungi Associated with Collar Rot on the Severity of the Disease :

Data presented in Table 9 showed that maximum lesion length (20.00 mm) was produced on rice plants inoculated with *P. versicolor* and 6.75 mm lesion by *C. globosum* respectively. Fungi when inoculated in combination produced lesser lesion length than those produced by *P. versicolor* but greater than *C. globosum.*

4.4.2 Effect of First Collar, First and Second and First, Second and Third Affected Collars when Inoculated with either *P. versicolor* of *C. globosum* on Rice Yield :

It was appeared from Table 10 that Leima Phou (KD, 2-6-3) rice plant when affected first, second and third collar when inoculated with either *P. versicolor* or *C. globosum* gave lowest yield of 4635.0 and 4701.5 kg than the rice plants affected only first collar (5286.2 and 5396.3 kg) and first and second collars (4901.5 and

4985.0 Kg)/ha respectively.

TABLE - 9

Effect of Infection Singly or in Combination of Fungi Associated with Collar Rot on the Severity of the Disease

Sl.No.	Inoculated Organism Singly	Combination	Rice variety	Lession length
1.	*P. versicolor*	-	KD. 2-6-3	Leaf blade 9.0mm + Leaf sheath 11.0mm --------- 20.3mm.
2.	*C. globosum*		- do -	Leaf blade 3.25 mm. + Leaf sheath 3.50 mm. ---------- 6.75 mm.
3.	-	*P. versicolor* + *C. globosum*	- do -	Leaf blade 3.0 mm. + Leaf sheat 6.3 mm. ---------- 9.3 mm.

4.4.3 Effect of Artificial Inoculation of Collar Rot Fungi on Rice Yield:

When the fungi were artificially inoculated separately on three rice varieties namely, Taichung Native-1, Leima Phou (KD, 2-6-3) and China dwarf at tillering and booting stages produced different disease reactions (Table 11). It was observed that all the varieties inoculated with *P. versicolor* at tillering stage showed cent per cent disease, which further, resulted in death of the rice plants. All the varieties inoculated with both the fungi (*P.versicolor* and *C. globosum*) at tillering stage were found to be more severe as compared to those inoculated with these fungi at booting stage. It was further observed that Taichung Native-1 was highly susceptible when inoculated with these fungi at both stages resulting in the reduction

TABLE - 10

Effect of First Collar, First and Second and First, Second and Third Affected Collars when Inoculated with either *P. Versicolor* or *C. Globosum* on Rice Yield

S.No.	Pathogen	Rice variety	Stage	Inoculation No. of collar	Weight 1000 grains (g*)	Yield Kg/.ha	Yield Reduction (%)
1.	*P. versicolor*	KD, 2-6-3	Booting	1 collar affected	21.74	5286.2	24.82
				2 collar affected	20.65	4901.5	30.00
				3 collar affected	19.95	4635.0	34.07
2.	*C. globosum*	KD, 2-6-3	Booting	1 collar affected	22.54	5396.3	23.25
				2 collar affected	21.98	4985.0	29.00
				3 collar affected	21.00	4701.5	33.13
3.		KD, 2-6-3		Control		7031.0	

* Average of 3 replications

TABLE - 11

Effect of Artificial Inoculation of Collar Rot Fungi on Rice Yield

S.No.	Variety	Stage of inoculation				Severity of disease (%)				Yield 1000 grains (g)			
		Tillering Stage A		Boot Stage B		A		B		A		B	
		P.v.	*C.g.*	*P.v.*	*C.g.*	*P.v.*	*C.g.*	*P.v.*	*C.g.*	*P.v.*	*C.g.*	*P.v.*	*C.g.*
1.	KD 2-6-3	25	25	25	25	100	75	97	60	0	22.2	20.43	21.69
	Control	25	25	25	25	0	0	0	0	28.5	28.5	30.50	30.50
2.	Taichung	25	25	25	25	100	80	90	32	0	15.6	13.30	16.00
	Control	25	25	25	25	0	0	0	0	23.7	23.7	24.00	24.00
3.	China dwarf	25	25	25	25	100	81	95	36	0	20.1	17.00	19.4
	Control	25	25	25	25	0	0	0	0	25.5	25.5	27.50	27.50

P.v. = *Pestalotiopsis versicolor*
C.g. = *Chaetomium globosum*

of yield/ 1,000 grain followed by China dwarf and Leima Phou (KD, 2-6-3).

4.5. BIOCHEMICAL CHANGES IN THE COLLAR INFECTED WITH COLLAR ROT FUNGI (TOTAL SOLUBLE SUGARS, REDUCING SUGARS, NON-REDUCING SUGARS AND FREE AMINO ACIDS:

Eight ninhydrin positive compounds namely, alanine, arginine, cystine, histidine, phenylalanine, methionine, tyrosin and valine were recovered from healthy and diseased collars. Maximum quantity (0.13 mg/g) was contributed by agrinine followed by methionine, histidine, valine and alanine respectively (Table 12).

The concentration of total free amino acids were increased considerably in diseased collars caused by both fungi as compared to healthy (Table 12). However, there was no qualitative change in free amino acid content of healthy and diseased collars which were affected by *C.globosum*. But valine could not be recovered from rice collars which were affected by *P. versicolor*.

The quantity of total soluble sugars and reducing sugars was increased in diseased collars (Table 12) due to infection with either *P. versicolor* or *C. globosum*.

The quantity of total soluble sugars and reducing sugars was increased in diseased collars (Table 12) due to infection with either *P. versicolor* or C. *globosum*. However, the concentration of non reducing sugar decreased in diseased tissues affected by either *P. versicolor* or *C. globosum* respectively.

4.6. SCREENING OF THE AVAILABLE HIGH YIELDING AND LOCAL RICE CULTIVARS OF MANIPUR AGAINST COLLAR ROT:

Out of 168 rice cultivars/varieties, 163 cultivars/ varieties were found to be susceptible against collar rot fungi. Sangsangba, Tumai,

Dullar, Tetep and CAU selection-I showed highly resistant against the disease (Table 13, 14 and 15).

TABLE - 12

Biochemical Changes in the Collar Infected with Collar Rot Fungi (Total Soluble Sugars, Reducing Sugars, Non-Reducing Sugars and Free Amino Acids)

Constituent	*P. versicolor* (mg)		*C. globosum* (mg)	
	Healthy	Diseased	Healthy	Diseased
A. Carbohydratres:				
Total soluble sugar	1.77	1.92	1.78	1.8
Reducing sugar	0.51	1.05	0.55	0.74
Non-Reducing sugar	1.19	0.79	1.23	0.70
B. Amino acids :				
Alanine	0.07	0.12	0.06	0.11
Arginine	0.13	0.15	0.13	0.16
Cystine	0.06	0.09	0.06	0.08
Histidine	0.11	0.14	0.12	0.15
Phenylalanine	0.08	0.17	0.08	0.16
Methionine	0.12	0.14	0.12	0.14
Tyrosine	0.06	0.09	0.06	0.09
Valine	0.09	0.00	0.08	0.12
C. Total free amino acids	0.72	0.90	0.71	1.01

TABLE - 13

Reaction of Local Rice Varieties Against Rice Collar Rot

Variety	Reaction
Auchadi, Bougreimanap, Jing-Sang-Lang, Maha-Phicaro-Mano and Xengan.	Susceptible **
Ahangrei, Ahaza, Akhan-phou, Akhut-sou, Akia, Ashama, Asham-Ju-Jaka, Asham-Ju, Ngao, Aso, Ate, Athebu, Badiro-Mowo, Banglei, Bilkuang - Buh, Buhman, Buhman-Sen, Buhman-Tiat, Buhman-Te, Guhman, Buhmui, Chadet, Chahao-Amubi, Chahao-Angouba, Chahao-poireiton, Chahao-shempak, Champhai, Champhai-Phoujao, Champi, Chanajhu, Chandrika, Changhil, Changlei, Chavummu, Chingphou-Akuppa, Ching-Phourel, Chitebuh, Dalai, Daziiro, Disha, Galbuh, Galleng-Buh-Pi, Ganesh-Phou, Haflong-Tei, Heimang-Jhu, Heitup-Phou, Housir, Huikap-Phou, Ijong, Iroya, Jaguding, Jingsangla, Kahaola, Kahara, Kaibu-Ziro, Kaijah, Kakchengphou, Kanglou, Kapo, Kaphungza, Kapongla, Kawng lawng, Khashok-Jhu, Khasia, Khongman, Khuangman, Maziiro, Mikro-Ro, Minri, Mojha, Muza, Nagarah, Nahazong, Napang-Kongkoi, Napang-Kong, Koi, Naphang-Duimei, Napi-Phou, Napok-Nap, Nga-Bupal, Naga-Buh, Nagarah, Ngamrung-Buh, Nahozhero, Noining, Oromono, Paocheijhu, Paoreima, Phe-Charang, Phourel, Pourel-Tujaombi, Sachi-Khong, Saituol Sang-Buh,	Susceptible * Susceptible*
C.A.U. Selection - 1, Sangsangba and Tumai	Resistant

Susceptible* : The whole collar region rotted.
Susceptible** : The whole collar region rotted and leaf dropped.

TABLE - 14

Reaction of Exotic Rice Varieties against Rice Collar Rot

Variety	Reaction
CT-C, CT-7, CT-112, CT-147 CT-225, IET-1159, IET-4095 IET-4576, IET-8682, IET-9689 IET-9691, IET-9692, IR-24, IR-30, IER-32, IR-38, IR-42, IR-50, IR-1561, IR-81185, Himalaya-2, Hira, Kalyani, Nila, RCM-8, Seoul and Tara	Susceptible*
Dullar and Tetep	Resistant

Susceptible* : The whole collar region rotted.

TABLE - 15

Reaction of IET Rice Lines Against Rice Collar Rot

IET NO.	Reaction
IET-13933, IET-14296, IET-14297,IET-14298, IET-14299, IET-14300, IET-14302, IET-14303, IET-14306, IET-14309, IET-14311, IET-14341, IET-14793, IET-14795, IET-14222, IET-14223, IET-14229, IET-14236, IET-14237, IET-14239 & IET-14241,	Susceptible**

IET-14314, IET-14791, IET-14803, IET-14805, IET-14808, IET-14807, IET-14814, IET-14817, IET-14219, IET-14220, IET-14252, IET-14253, IET-14254, IET-14725, IET-14726, IET-14727, IET-14728 & IET-14729	Susceptible*

Susceptible * : The whole collar region rotted.
Susceptible** : The whole collar region rotted and leaf dropped

TABLE - 16

Effect of Different Fungicides on Growth of Collar Rot Fungi

Sl. No.	Fungicide	conc. (%)	Growth (mg*)		Inhibition on growth (%)	
			P. v.	*C. g.*	*P. v.*	*C. g.*
1.	Bavistin	0.1	0.00	0.00	100	100
		0.05	0.00	0.00	100	100
		0.025	0.00	0.00	100	100
2.	Blitox-50	0.2	0.00	0.00	100	100
		0.05	174.50	0.00	68.77	100
		0.025	251.75	0.00	54.94	100
3.	Dithane M-45					
		0.2	0.00	0.00	100	100
		0.05	0.00	0.00	100	100
		0.025	0.00	0.00	100	100
4.	Foltaf	0.1	0.00	0.00	100	100
		0.05	0.00	0.00	100	100
		0.025	0.00	0.00	100	100
5.	Topsin-M	0.1	0.00	0.00	100	100
		0.05	0.00	0.00	100	100
		0.025	0.00	0.00	100	100
6.	Control	0.00	368.685	368.685	0.00	0

P.v. = *Pestalotiopsis versicolor*
Ch.g.= *Chaetomium globosum*

4.7.1 Effect of Different Fungicides on Growth and Sporulation of Rice Collar Rot Fungi:

It was observed from Table 16 that Bavistin, Blitox-50, Dithane M-45, Foltaf and Topsin-M could completely control the growth of *P. versicolor* and *C. globosum* at the recommended doses. However, at lower concentration, 0.025 per cent each of Bavistin, Dithane M-45, Foltaf and Topsin M, both fungi could not grow. But at low concentration (0.05 and 0.025) of Blitox, *C. globosum* could not grow but *P. versicolor* could grow.

TABLE - 17
Fungistatic and Fungicidal Effects of Fungicides on the Growth of Collar Rot Fungi

Sl. No.	Fungicides	Concentration %	Growth			
			Fungistatic		Fungicidal	
			P. versicolor	*C. globosum*	*P. versicolor*	*C. globosum*
1.	Bavistin	0.1	0	0	+	+
		0.05	0	0	+	+
		0.025	0	0	+	+
2.	Blitox-50	0.2	+	0	0	+
3.	Dithane M-45	0.2	0	0	+	+
		0.05	0	0	+	+
		0.025	0	0	+	+
4.	Topsin-M	0.1	0	0	+	+
		0.05	0	0	+	+
		0.025	0	0	+	+
5.	Foltaf	0.1	0	0	+	+
		0.05	0	0	+	+
		0.025	0	0	+	+

TABLE - 18

Effect of Fungicides on Disease Intensity and Rice Yield

Sl.No	Fungicides	Concentration (%)	Disease intensity (%)		Yield 1000 grain (g)	
			P. versicolor	*C. globosum*	*P. versicolor*	*C. globosum*
1.	Bavistin	0.1	0	0	30.63	30.63
2.	Blitox-50	0.2	80	0	15.75	28.30
3.	Dithane M-45	0.2	0	0	26.50	29.50
4.	Foltaf	0.1	0	0	27.25	30.10
5.	Topsin M	0.1	0	0	28.35	30.50
6.	Control	0.00	100	100	8.45	16.70

It was clearly seen from Table 17 that mycelial plugs of *P. versicolor* and *C. globosum* which could not grow on potato dextrose broth posioned with each fungicide were taken out after 7 days icubation and washed thrice in sterile distilled water could grow/not grow. Hence, Bavistin, Dithane M-45, Foltaf and Topsin-M showed fungicidal action to both fungi. However, Blitox-50 had fungistatic effect on these two fungi.

4.7.2 Effect of Fungicides on Disease Intensity and Rice Yield:

Results on Table 18 indicated that Bavistin, Dithane M-45, Foltaf and Topsin-M could control collar rot disease of rice with increased grain yield. However, Blitox-50 could control disease caused by *C. globosum* and resulting in increased grain yield, whereas rice plants infected with *P. versicolor* when sprayed with Blitox-50 showed 80 per cent disease intensity and gave lower yield (15.75 g/1000 grain wt.).

Chapter 5

Discussion

5.2 ISOLATION OF FUNGI ASSOCIATED WITH RICE COLLAR ROT:

P. versicolor (Speg.) and *C. globosum* were consistantly found associated with rice collar rot. KD. 2-6-3 and China dwarf rice plants artificially inoculated with the pure culture of both fungi produced the collar rot disease symptoms after 3 days of inoculation. Similarly, Singh *et al.* (1995) isolated *P. versiocolor* from the rice collar rot samples.

5.2.a. Distribution of Fungi Associated with Collar Rot of Rice at Various Localities in Different Districts of Manipur.

The fungi were well distributed at various localities in Imphal, Thoubal, Chandel, Bishnupur and Senapati districts of Manipur where H YV were grown. Present findings were in agreement with Amin *et al.* (1974) who reported that high yielding rice varieties were severely affected by *Acrocylindrium oryzae*. Similarly, Devi (1991) observed that high yielding rice varieties were more susceptible than the tall local varieties to sheath rot fungi namely, *F. graminearum; F.moniliforme* and *F. avenaceum*.

5.3.1. Effect of Media on the Growth and Sporulation of Collar Rot Fungi:

All the media under study supported mycelial growth of the collar rot fungi. Maximum mycelial growth of *P. versioclor* was found in Richard's medium followed by potato dextrose but least in Czapek-Dox. Similarly, Palarpaman (1987) reported that isolate 2 of *Colletotricum curcumae* showed maximum growth on Richards's medium and dextrose asparagine phosphate medium followed by Czapek Dox medium. Potato dextrose broth supported maximum mycelial growth of *C. globosum* followed by Corn meal but least in Czapek Dox. Similarly, potato dextrose broth was found to be best for the growth of *Acrocylindrium oryzae* (Tasugi and Ikeda, 1956) and *Sarocladium oryzae* (Chin and Thseng, 1981).

Devi (1991) observed the maximum mycelial growth of three fusaria sheath rot fungi in potato dextrose broth followed by host extract and Richard's medium.

5.3.2. Effect of pH on the Growth and Sporulation of Collar Rot Fungi:

A good fungal growth of *P. versicolor* and *C. globosum* were obtained in a wide range of pH. However, maximum mycelial growth of *P. versicolor* was at pH 5 and *C. globosum* was at pH 8. Similarly pH 5 was found maximum for the growth of *Macrophomina phaseoli* of groundnut and *P. versicolor* Nogdren khomba (1993). Further it was reported that a wide range pH i.e. 5.0-8.0 could support the mycelial growth of *Sarocladium oryzae* with optimum pH at 5.5-6.0 (Datta and Purkayastha, 1978). However, *P. versicolor* and *C. globosum* failed to sporulate in all pH values tested upto 7 days of incubation. Similar observations have been made by Nongdren Khomba (1993) who found that *P. versicolor* failed to sporulate upto 7 days.

5.3.3.a. Effect of Different Carbon Sources on the Growth and Sporulation of Collar Rot Fungi :

Maximum mycelial growth of *P. versicolor* was found in cellobiose but least in ribose. However, there was not singinficantly different in the mycelial growth of this fungus in mannose, cellobiose, maltose, fructose and mannitol respectively. And maximum mycelial growth of *C. globosum* was found in mannose followed by sucrose least in arabinose. However, there was not significantly different in mycelial growth of this fungus in all the test carbon sources. Das *et al.* (1985) observed sucrose as the best carbohydrate which significantly promoted mycelial growth of *Pestalotia palmorum* followed by mannitol, glucose, lactose, frcutose and starch.

Devi (1991) reported that maximum mycelial growth of fusaria sheath rot in sucrose, D(+) Xylose and D-fructose but lower growth was obtained in Glycerol and (-) Arabinose repectively. It was also observed that carbon sources under test failed to induce sporulation of both fungi upto 7 days of incubation.

5.3.3.b. Effect of Amino Acids on the Growth and Sporulation of Collar Rot Fungi:

While growing the collar rot fungi namely *P. versicolor* and *C. globosum* in medium containing each amino acid under test, it was observed that L-lycine supported maximum mycelial growth of *P. versicolor* followed by DL-aspartic acid but poor in L-histionine, L-leucine supported maximum mycelial growth of *C. globosum* followed by L-lycine but least in DL-methionine. Wills (1968) found that DL-glutamic acid and DL-aspartic acid supportd good growth of the *Sclerotinia sclerotium* isolate SL-4 and K-3 whereas L-metheonine, L-cystein and cystine were poor sources of nitrogen. Similarly Wang and Tourneau (1972) found that aspartic acid, glutamic acid, serine and alanine were superior nitrogen sources for growth and sclerotia formation of *sclerotinia sclerotiums*.

Devi (1991) reported that L-histidine, DL-aspartic acid and DL-9-alanine supported maximum mycelial growth of *Fusarium* sheath rot fungi. Further, observed that amino acids under test could not induce sporulation of both fungi upto 7-days of incubation.

5.3.4 Effect of Light on Growth and Sporulation of Collar Rot Fungi :

The three different conditions of light showed different effect on growth and sporulation of *P. versicolor* and *C. globosum*. The maximum mycelial growth of both fungi was found in complete darkness but least in alternate light and darkness. Zhuge and Lapis (1985) reported that continuous darkness favoured good growth and sporulation of *Sarocladium oryzae*. Similar observations were made by Devi and Singh (1994) that *Fusarium avenaceum* and *Fusarium graminearum* required complete darkness for their maximum mycelial growth.

In the case of sporulation of *P. versicolor*, complete light could induce sporulation after 4 days of incubation. However, *C. globosum* failed to sporulate in this condition. And also alternate light and darkness and complete darkness failed to induce sporulation of both fungi. The present findings are in agreement with Mandal and Chaudhuri (1990) who reported that direct sunlight and diffused light conditions favoured chlaymydospore formation by *Fusarium moniliforme*. Similarly, Devi and Singh (1994) observed that *F. moniliforme* could induce maximum sporulation in continuous light.

5.3.5. Effect of Temperature on Growth and Sporulation of Fungi Associated with Collar Rot of Rice:

The mycelial growth of *P. versicolor* and *C. globosum* was favoured by a wide temperature range of 10-35°C. Optimum temperature for maximum mycelial growth of both fungi was recorded as 25°C. However, they fail to grow at 5.0 and 38°C.

The present findings are in agreement with those of Kawamura (1940) and Zhuge and Lapis (1985) who found that rice sheath rot pathogen grown at 15-35°C with optimum at 30°C. In case of *Fusarium moniliforme*, the optimum was about 27-30°C and minimum 7-8°C (Hemmi and Aoyagi, 1941). Similarly, Devi and Singh (1994) reported that the growth of *F. avenaceum*, *F. moniliforme* and *F. graminearum* were favoured by a wide temperature range of 10-35°C Optimum temperature for maximum mycelial growth of *F. avenaceum* was found at 25°C and that of *F. moniliforme* and *F. graminearum* at 30°C. But they failed to grow at 5 and 38°C. respectively.

However, *P. versicolor* and *C. globosum* did not sporulate at all temperatures studied upto 7 days of incubation. The present findings are in agreement with those of Devi and Singh (1994) who found that temperatures 5, 10, 15, 35 and 38°C were unfavourable for sporulation of *F. avenaceum*, *F. moniliforme* and *F. graminearum*.

5.3.6. Effect of Relative Humidity (R.H.) on the Germination of Fungi Associated with Rice Collar Rot:

Spores of *P. versicolor* and *C. globosum* started germination at 50 per cent relative humidity *P. versicolor* germinated after 35 hours of incubation from 50-100 per cent relative humidity and those of *C. globosum* at 50-100 per cent relative humidity after 24 hours of incubation with different per cent of germination. Maximum spore germination of both fungi was recorded to be 100 per cent at 100 per cent relative humidity. It was also observed that at low relative humidity, per cent of germination of both fungi was lower, whereas at increasing R.H., per cent of germination was increased.

Similarly, Inoue and Uchino (1963) reported that high relative humidity above 95-96 per cent was required for sclerotial germination of rice sheath blight. Critical air humidity for spore germination of *P. oryzae* was about 92 and 96 per cent (Abe, 1933 b.).

Similar observation was made by Devi (1991) who reported that *F. avenaceum* and *F. moniliforme* started germination of relative humidity of 40 and 60 per cent after 24 hours of incubation and *F. graminearum* at relative humidity of 50 per cent after 40 hours of incubation. Maximum spore germination of *F. avenaceum*, *F. moniliforme* and *F. graminearum* was recorded to be 96, 93 abd

92 per cent respectively at 75, 100 and 75 per cent relative humidity.

5.4.1. Effect of Infection Singly or in Combination of Fungi Associated with Collar Rot on the Severity of Rice Collar Rot :

P. versicolor and *C. globosum* when inoculated artificially to Leima Phou (K.D.2-6-3) rice plant produced typical symptoms of rice collar rot. The length of diseased lesions was more in rice plants which were inoculated singly with each of the fungus as compared to that of inoculated with in combinations. Mix inoculation showed marked antagonistic effect as the length of disease lesion was reduced.

The antagonistic effect may be due to the production of toxic metabolites by *P. versicolor and C. globosum* and interact themselves. The production of toxic metabolites by *A. oryzae* and *F. moniliforme in vitro* has been reported by various workers (Chung, 1975 and Thankur, 1973).

5.4.2. Effect of First Collar, First and Second and First, Second and Third Affected Collars when Inoculated with either *P. versicolor* or *C. globosum* on Rice Yield:

First, second and third collars infected rice plants with either *P. versiocolor* of *C. globosum* gave highest reduction in yield followed by first and second affected collars but least reduction in first collar affected rice plant. It was evident that rice yield increased with less number of collar affected with collar rot fungi. Hence yield damage was found to be highly correlated with increased in collar affected by both fungi.

5.4.3. Effect of artificial inoculation of collar rot fungi on rice yield :

Three rice varieties namely, Taichung Native-1, Leima Phou

(KD. 2-6-3) and China dwarf showed susceptible reactions when inoculated with the collar rot fungi at tillering and booting stage resulting in marked reduction on 1,000 grain weight. Similarly Tikoo (1985) reported that the number of spikelets/panicle; number of grains/ panicle and 1,000 grain weight were reduced markedly due to sheath rot incidence on Cu-k-78-13.

5.5. BIOCHEMICAL CHANGES IN COLLAR INFECTED WITH COLLAR ROT FUNGI (TOTAL SOLUBLE SUGARS, REDUCING SUGARS, NON REDUCING SUGARS AND FREE AMINO ACIDS) :

Eight ninhydrin positive compounds namely, alanine, arginine, cystine, histidine, phenylalanine, methionine, tyrosine and valine were recovered from healthy and diseased collars. Similarly, Yadav *et al.* (1986) recovered many amino acids from diseased and healthy leaves. Maximum quantity (0.13 mg/g) was contributed by arginine followed by methionine histidine, alanine and phenylalanine respectively.

The concentration of total free amino acids was increased considerably in infected collars caused by both fungi as compared to healthy collars. However, no qualitative change in free amino acids was found in healthy and diseased collars caused by *C. globosum*. But diseased collars caused by *P. versicolor* had slight qualitative change in free amino acids content i.e. valine, could not be recovered from rice collar affected by *P. versicolor*. The present findings are in agreement with those of Pundir *et al.* (1990) who reported that concentration of free amino acids was more in infected lentil leaves and there was no qualitative change in infected as well as healthy leaves. The increase in total free amino acid in diseased rice collars could be due to the degradation of host proteins into free amino acids by fungal proteases.

The quantity of total soluble sugars, and non reducing sugars was increased in diseased collar due to infection with either *P. versicolor* or *C. globosum*. The present findings are in accordance with those of Iboton *et al.* (1982) who found that concentration of reducing sugars increase in greening infected seedlings. Similar observations have also been made by Parsad *et al.* (1990) who reported increase in reducing sugars in storage seeds of raddish infected with *Aspergillus flavous*. The increase in reducing sugars may be due to hydrolysis of starch and cellulose by amylase and cellulase respectively.

However, the concentration of non-reducing sugar decreased in diseased tissues infected by either *P. versicolor* or *C. globosom* respectively.

5.6. SCREENING OF THE AVAILABLE HIGH YIELDING AND LOCAL RICE CULTIVARS OF MANIPUR AGAINST COLLAR ROT:

Results of varietal screening against collar rot pathogens, showed resistant to susceptible reaction. Out of 168 rice cultivars/varieties field tested in two consecutive crop seasons including high yielding varieties, local rice cultivars and improved varieties inoculated with collar rot fungi. 163 varieties/cultivars were found to be susceptible. Two local rice varieties (Sangsangba and Tumai) have been found resistant and among the exotic and improved varieties Dullar, Tetep and CAU selection-I showed resistant to collar rot fungi.

Similarly varietal differences in susceptibility of rice plants to sheath rot in Taiwan were observed by Chen and Chien (1964). The reactions of some indigenous germplasm of released and improved rice cultivars to sheath rot were scored as highly resistant and highly susceptible by Subramanian and Ramakrishnan (1975); Amin (1976); Shahjahan *et al.* (1977) and Murkerjee *et al.* (1981). Chien and Thseng (1982), Lewin and Vidhyasekaran (1987), Grewal and Kang (1988) and Devi (1991) under field and glass house conditions.

This seems to conclude that differences in susceptibility and resistance may be due to different genetic make up of the rice plants and different fungi envolved.

5.7.1. Effect of Different Fungicides on Growth and Sporulation of Rice Collar Rot Fungi :

Of the five fungicides tested *in vitro* , *P. versicolor* and *C. globosum* failed to grow in Bavistin, Dithane M-45, Foltaf and Tospin-M at all concentrations. And also all concentrations of Blitox-50 inhibited completely the growth of *C. globosum.* But the fungus. *P. versicolor* could grow well at 0.05 and 0.025 concentrations of Blitox-50 and the percentages of inhibition over control were 68.77 and 54.94 at 0.05 and 0.025 per cent. But Blitox at 0.2 per cent could inhibit completely the fungal growth.

Similarly. Reddy *et al.* (1984) found that *Acrocylindrium oryzae* was significantly inhibited by Bavistin followed by Blue Copper and Antracol *in vitro.* Devi (1991) also reported that three fusaria sheath rot fungi could be effectively controlled by Bavistin, Benlate and Dithane M-45 *in vitro.*

In vitro effectiveness of Topsin-M and Dithane M-45 had also been demonstrated by Akbari *et al.* (1992). Upadhyay, (1984) found that in *in vitro* tests phygon XL (dichlone) was most effective against *Pestalotiopsis funerea* (a causal agent of *Eucalyptus* leaf spot) followed by Difolatan 80 W (Captafol), Dithane Z-78 (Zineb), Zincop and Cupramar (Copper oxychloride).

5.7.1.a. Fungistatic and Fungicidal Effects of Fungicides on the Growth of Collar Rot Fungi:

After testing in the laboratory, Bavistin, Dithane M-45, Foltaf and Topsin-M showed fungicidal action to *P. versicolor* and *C. globosum.*However, Blitox-50 had fungistatic effect on these two fungi.

5.7.2. Effect of Fungicides on Disease Intensity and Rice Yield:

Bavistin, Dithane M-45, Foltaf and Topsin-M could control the disease under field condition with increased grain yield. However, Blitox-50 also could control disease caused by *C. globosum* with increased grain yield but failed to check the disease caused by *P. versicolor* which produced 80 per cent of disease intensity and gave lower yield. These findings corroborate those of Sugha and Singh (1989) who observed the effectiveness of Bavistin, Mancozeb, Busan and Benlate. Similar observations were made by Devi (1991) who reported that Dithane M-45, Bavistin and Benlate could reduce disease intensity resulting in increased grain yield.

Chapter 6

Summary

The present investigations were carried out in the Department of Plant Pathology, College of Agricultural, Central Agricultural University, Iroisemba, Imphal. Various experiments were conducted to gain insight into different aspects of rice collar rot disease, like effect of different media on the growth and sporulation; effect of carbon nitrogen sources on the growth and sporulation of collar rot fungi; effect of artificial inoculation either singly or incombination of fungi associated with collar rot on the severity of the disease; efficacy of different fungicides in controlling rice collar rot; effect of light, temperature and relative humidity on the germination of collar rot fungi; estimation of total soluble sugars, reducing sugars, non reducing sugars and free amino acids from diseased and healthy rice collars; screening of available high yielding and local rice cultivars against rice collar rot. The salient features of the investigations are summerised below.

1. Two fungi, namely *Pestalotiopsis versicolor* (Speg.) and *Chaetomium globosum* were consistantly found to be associated with rice collar rot in Manipur. They were found well distributed at various rice growing localities in Imphal, Chandel, Senapati, Thoubal and Bishnupur Districts of Manipur. Both high yielding varieties and local cultivars of rice were found to be severely effected by these fungi.

2. Of the seven different types of culture media tested for growth and sporulation of fungi associated with collar rot, Richard broth supported maximum mycelial growth of *P. versicolor* (Speg.) and

Potato Dextrose broth supported maximum mycelial growth of *C. globosum*. No sporulation of the fungi was found in all culture media upto 7 days of incubation.

3. The fungi associated with rice collar rot could grow in a wide range of pH. However, the maximum mycelial growth of *P. versicolor* was found in pH 5 and *C. globosum* was found in pH 8. But both the fungi failed to grow at pH 12. No sporulation of fungi was found in all pH values tested upto 7 days of incubation.

4. Of the different carbon sources tested for growth and sporulation of fungi, cellobiose supported maximum mycelial growth of *P. versicolor* and mannose supported maximum mycelial growth of *C. globosum*. None of the carbon sources support sporulation of these fungi upto 7 days of incubation. However, both fungi produced their spores after 15 days of incubation.

5. Of the twentyone amino acids, L-Lucine supported maximum myclial growth of *P. versicolor* followed by DL-Aspertic acid. L-Leucine supported maximum mycelial growth of *C. globosum* followed by L-Lucine. Amino acids failed to induce sporulation upto 7 days of incubation. However both the fungi sporulate after 15 days of incubation.

6. Light played an important role in growth and sporulation of collar rot fungi. *P. versicolor* required complete darkness for their maximum mycelial growth and also found the same with *C. globosum*. Maximum sporulation of *P. versicolor* was induced by complete light. Sporulation of the *C. globosum* could not be induced by light and darkness upto 7 days of incubation. Sporulation of both fungi occurred after 15 days of incubation.

7. Temperature played an important role in growth of collar rot fungi. The growth of collar rot fungi was favoured by a wide temperature range, *P. versicolor* and *C. globosum* showed an optimum temperature of 25°C for maximum growth.

8. *P. versicolor* started germination at 50 per cent relative humidity after 35 hours of incubation and *C. globosum* at 50 per cent relative humidity after 24 hours of incubation. Maximum spore germination of both fungi was recorded to be 100 per cent at 100 per cent realtive humidity. The per cent of germination of both fungi increased with increasing relative humidity and decreased with lowering relative humidity respectively.

9. The length of diseased lesions (on collar and leaf sheath) was more in Leima Phou (KD, 2-6-3) rice plants which were inoculated with *P. versicolor* of *C. globosum* singly as compared to those inoculated in combinations. The combination of *P. versicolor* + *C.globosum* showed antagonistic effect as minimum lesion length was produced. Yield was reduced upto about 35% over control under *control* conditions.

10. Collars from diseased and healthy plants were analysed for biochemical constituents. The concentration of free amino acids was increased in diseased collars caused by both fungi. Eight ninhydrin positive compounds, amino acids were recovered from both healthy and diseased collars. The amount of total soluble sugars, reducing sugars was increased in rice collar infected either with *P. versicolor* or *C. globosum*. However, the concentration of non-reducing sugars decreased in diseased tissues associated with either *P. versicolor* or *C. golobosum*.

11. Of 168 rice cultivars/ varieties field tested including high yielding varieties, local rice cultivars and improved varieties inoculated with collar rot fungi, 163 varieties/ cultivars were found to be susceptible. Sangsangba, Tumai, Dullar, Tetep and C A U Selection-I were found resistant.

12. Of the fungicides tested, Bavistin, Dithane M-45, Foltaf and Topsin-M could completely check the growth of *P. versicolor* and *C. globosum*.

While plated the fungal plugs treated with either Bavistin or

Dithane M-45 or Foltaf or Topsin-M on PDA, fungicides were found fungicidal whereas Blitox-50 proved fungistatic to the fungus *P. versicolor* only.

13. Of the chemicals tested Dithane M-45, Folta, Topsin-M and Bavistin could control the disease caused by *P. versicolor* and *C. globosum* resulting in increased grain yields under field conditions.

Bibliography

Abe, T. 1933b : The relationship on atmospheric humidity to infection of the rice plants by *Pyricularia oryzae*. *Review of Applied Mycology*, **13**: 265.

Agarwal, G.P. and Ganguli S. (1960) : The influence of different sources of carbon on the growth and sporulation of *Pestalotiopsis versicolor* (Speg.) Stey, *Proc. Nat. Acad Sci.* India **30**; 51-54.

Agarwala, R.K. 1955 : D. Phil. Thesis, University of Allahabad.

Akbari. P. P., Rawal and Khandar, R.R. 1992 : Evaluation of fungicides against *Cercospora .citrullina* cooke. Causing leaf spot of bottle gourd. *Indian J. Mycol. Pl. Pathol.*, **22** (1) : 89.

Amin, K.S. 1976 : Source of resistance to *Acrocylindrium* sheath rot of rice. *Plant Dis. Reptr.*, **60**: 72-73.

Amin, K.S., Sharma B. D. and Das, C.R. 1974 : Occurrence in India of sheath rot of rice caused by *Acrocylindrium*. *Plant Dis. Reptr.*, **58** (40) : 358-360.

Bakr, M.A. and Miah, S.A. 1975 : Leaf scald of rice, a new disease in Bangladesh. *Plant Dis. Reptr.*, **55**: 909.

Bhargava, S.N. (1962) : D. Phil. Thesis, University of Allahabad.

Chakravarty, D.K. and Biswas, S. 1978. Estimation of yield loss in rice affected by sheath rot. *Plant Dis. Reptr.*, ***62***: 226-227.

Chandra, S. 1961 : D. Phil. Thesis, University of Allahabad.

Chen, C.C. and Chien, C.C. 1964 : Some observations on the out break of rice sheath rot disease. *J. Taiwan Agric. Res.*, **13**: 39-45.

Chien, C.C. and Thseng, F.M. 1981 : Physiological and biological studies of rice sheath rot pathogen, *J. Agric. Res.* China, **30**: 1-4.

Chien, C.C. and Thseng, F.M. 1982 : Rice varieties and in relation to the susceptibility to sheath rot. *J. Agric. Res.* China, **31**: (2) : 155-161.

Chin, K.M. 1974 : Collar rot, a new symptom of the rice blast disease (*Pyricularia oryzae*) MARDI-Res-Bull-Malays-Agric-Res-Dev-Inst., **2** (2) : 82-84

Chinnaswamy, R. Nair, M.C. and Menon, M.R. 1981 : Comparative efficacy of certain fungicides in the control of sheath rot of rice. *Agricultural Research Journal Kerala*, **19**: (2), 86-88.

Chung, H.S. 1975 : Studies on sheath rot of rice caused by *A. oryzae* Sawada-testing varietal reaction and culture filtrates of the causal fungus. *Korean J. Pl. Protection*, **14**: 23-27.

Das, C.M. and Mohanta, I.C. 1985 : Evaluation of some fungicides against *Pestalotia palmarum* Cke., incitant of grey blight of coconut. *Pesticides.*, **19**: 5, 37-39.

Das, C.M.; Mishra, S.K. and Swain, P. 1985 : Effect of different sources of carbon and nitrogen on the growth and sporulation of *Pestalotia palmarum* Cke. causing grey blight of coconut. *Indian J. Mycol. Pl. Path.*, **15**: (3): 325-327.

Datta, S.K. and Purkayastha, R.P. 1978 : Germination behaviour of *Acrocylindrium oryzae* causing sheath rot of rice. *Indian Phytopath.*, **31**: (1) : 89-91.

Devi, R.K. Tombisana. 1991 : Studies on fungi associated with rice sheath rot. Ph. D. thesis , Manipur University 117 pp.

Devi, R. K. Tombisana and Iboton, S.N. 1994 : Effect of temperature and light on growth and sporulation of fusarium rice sheath rot. *Intern. Rice Res. Notes.*, **19** (3) : 28.

Dhindra, H.S. Aaulak, K.S. and Chahal, S.S. 1991 : Incidence and assessment of losses due to false smut of rice in Punjab. *Indian Phytopath.*, **44**: 120-121.

Dube, H.C. 1966 : D. Phil. Thesis, University of Allahabad.

Eastrada, B.A.; Sanchez, L.M. and Crill, P. 1979 : Evaluation of screening methods for sheath rot resistance of rice (*Oryza sativa*). *Plant Dis. Reptr.*, **63**: 908-911.

Gaerber, U. and Motte, G. 1990 : First results of evaluation of fungicides against *Ascochyta fabae* Speg. by means of foliar method Nachrichtenblatt-des-Deutschen-Pflanzens-Chutzdienstes (Germany, FDR)., **42** (11) : 171-173.

Gill, M.A.; Yasin, S.I.; Khan, T.Z.; Ahsanullah, Q. and Butt, M.A. 1993 : Evaluation of resistance to bakanae and foot rot disease. *International Rice Research Notes* (Philippines), **18** (3) : 14-15.

Grewal, S.K. and Kang, M. S : 1988: Screening rice germ plasm against *Fusarium* sheath rot (ShR) disease. *Intern. Rice Res. Newsl.*, **13** (3) : 14.

Gupta, O.P., Nauriyal J.P. and Kuorr L.C. 1972 : Citrus greening in the Indian Punjab. pp. 14-18. In W.C. price (ed.). *Proc. Fifth cout. Intern. Organiaztion citrus Virol.*, University Florida Press, Gainesville.

Hara, K. 1959 : A monograph of rice diseases. 3rd ed. Japan, Tokyo (Ja).

Hashioka, Y. and Makino, M. 1956 : Relation of nitrogen nutrition on rice plants to the susceptibility to four foliage diseases., *Res. Bull. Fac. Agric. Gifu. Univ.*, **6**: 58-66.

Hemmi, T. and Aoyagi, Z. 1941 : Ecological studies on important fungi pathogenic to the crops in the Far East. I. Germination of the macroconidia of *Gibberella fujikuroi* in relation to some environmental factors. *Ann. Phytopathol. Soc. Japan.*, **11**: 66-80. (Ja, En).

Hosokawa, T. and Kubota, H. 1957 : Maintenance of relative humidity by *hydrogen* sulphuric acid Ecology., **45**: 579-580.

Iboton, Kapur S.P. and Cheema, S.S. 1982 : Some metabolic changes in greening-infected grape fruit twigs. *J. Res. Agric. Univ.*, **19**: 227-29.

Iqbal, S.M.; Bakhsh, A. and Malik, R.A. 1990 : Identification of resistant sources to *Ascochyta* blight in lentil (*Lens culinaris*). LENS-Newsletter (ICA RDA). Lentil Experimental News Service., **17** (1) : 26-27.

IRRI. 1980 : Standard Evaluation system for Rice., 44 pp.

Kanjanasoon, P. 1962 : A note on a new leaf disease of rice in Thailand. *International Rice Commission Newsletter.*, **11** (3):22.

Kanjanasoon, P. 1963 : Pycnidia formation of the collar rot fungus of rice on different media. International Rice Commission Newsletter., **12** (2) : 8-10.

Kanjanasoon, P. 1965 : Studies on the bakanae disease of rice in Thailand. Doc. Agric. Thesis, Tokyo University, Japan.

Kapoor, S.; Singh, G. Gill, A.S. and Singh, B. 1990 : Lentil lives resistant to *Ascochyta* blight (*Lens culinaris*). LENS-*Newsletter* (ICARDA). Lentil Experimental News Service., **7** (2) : 26-28.

Kapur, S.P. Singh, N. I. and Cheema, S.S. 1984 : Capillary tube inoculation on efficient method for inoculating citrus trees with die-back pathogen. PAU. *J. Res.*, **21** (1) : 137-138.

Kawamura, E. 1940 : Notes on the sheath rot of rice plants with special reference to its causal organism, *Acrocylindrium oryzae* Saw. *Ann. Phytopathol. Soc. Japan.*, **10** : 55-59.

Ketiku, A.O. 1973 : Chemical composition of unripe (green) and ripe plantain (*Musa parastisiaca). J. Sci. Food Agric.*, **24**: 703-707.

Kurosawa, E. 1929 : On the cultural characters of the 'bakanae' disease fungi on various nutrient media and the temperature of their development. *Report of the Natural History Society of Formosa.*, **19**: 150-179 (Ja).

Kwon, S. H.; Song, H.S. Sonn, C.Y. and Yamakuchi, T. 1973 : Identification of *Fusarium* leaf spot (*Fusarium nivale*) newly reported in Korea. *Korean J. Plant Protection.*, **12**: 121-124.

Lal, B. and Tandon R.N. 1964 : Utilization of carbon sources by *Pestalotia bicolor* Ell. and Ev. causing leaf spot of *Diosphyros peregrina* (Gaev.) Gurke. *Proc. Nat. Acad. Sci.* (India), **34**: 463-466.

Lewin, H.D. and Vidhyasekaran, P : 1987 Sources of resistance to sheath rot (ShR). *Intern. Rice Res. Newsl.*, **12** (1) : 7.

Lilly, V.G. and Barnett H.L. 1951 : Physiology of Fungi. McGraw Hill/Book Company, New York, 7 pp.

Mahadevan, A. 1974 : Methods in physiological plant pathology. Sirabami Publications, Madras, 82 pp.

Mandal, D.N. and Chaudhuri, S. 1990 : Induction of chlamydospore in *Fusarium moniliforme Indian Phytopath.*, **43**: 420-426.

Mohan, R. and Subramanian, C.L. 1977: Influence of nitrogen nutrition and sheath rot disease on sugar content of two paddyvarieties. *Curr. Sci.*, **46**: 753-755.

Mukerjee, P.; Singh, B.B. and Rahman, F. 1981 : Testing of indigenous germplasm of rice against sheath rot by artificial inoculation. *Indian Phytopath.*, **43** (3) : 287-290.

Naik, M.K., Hirenath, P.C. and Hirenath, S.V. 1988 : Postinfectional changes in the betelvine leaves infected with *Colletotrichum gloeosporioides*. *Indian Phytopath.*, **41** (3) : 370-372.

Naito, H. and Koshimizu, Y. 1974 : Morphology and hyphal growth temperature of brown leaf spot fungus of rice (rice scald fungus). *Ann. Phytopathol. Soc. Japan.*, **40**; 319-328 (Ja En.).

Nelson, N. 1944 : A Photometric adaptation of Somogyi method for the determination of glucose. *J. Biol. Chem.*,

153: 375-380.

Nongdren Khomba S.L. 1993 : Studies on *Pestalotipsis versicolor* (Speg.) a causal agent of rice collar rot. M.Sc. Thesis submitted to Manipur University, 60 pp.

Ou, S.H. 1985 : Rice Diseases. 2nd ed. CAB. London, 380 pp.

Ou, S.H., Nuque, F.L. and Vergel de Dios, T.I. 1978 : Perfect stage of *Rhynchosporium oryzae* and symptoms of rice leaf scald disease. *Plant Dis. Reptr.*, **62**: 524-528.

Palarpamar, M.Y. 1987 : Growth and sporulation of *Colletotrichum capsici* and *Colletotrichum curcumae* on different culture media. *Indian J. Mycol. Pl. Pathol.*, **17** (2) : 208.

Pater, J.S., M. N. Kamat and Hingorani G.M. 1951 : *Pestalotia psidii* on Guava. *Indian Phytopath.*, **3**: 165-176.

Prabakaran, J.,Ragunathan, V. and Prasad, N.N. 1974 : Occurrence of Sheath rot disease of rice caused by *Acrocylindrium oryzae* Sawada. Annamalai Univ. *Agric. Res. Annual*, **4/5**: 182-183.

Prasad, B. K. Sao, R.N.; Narayan, N.; Singh, R.N.; Rahman, A; Singh, S.P.; Dayal, S. and Shahker, U: 1990. Changes in sugar and amino acid contents in stored radish seed due to *Aspergillus flavus*. *Indian Phytopath.*, **43**: 457-460.

Prasad, S.S. 1963 : D.Phil. Thesis, Bihar University.

Pundir, C.S.,Singh, Om. and Verma, H.C. 1990 : Effects of Fusarium wilt on free amino acids in the progeny of healthy and infected lentil. *Indian Phytopath.*, **43**: 580-582.

Punithalingam, E. 1980 : *Phomopis oryzae - sativae*. CMI. Description of Pathogenic Fungi and Bacteria No. 665, 2 pp.

Punithalingam, E. and Sharma, N. D. 1979 : New microfungi from South East Asia. *Nova Hedwigia*., **32** : 881-857.

Purohit, D.K. 1972 : Effect of different carbon and nitrogen sources on sporulation of some pathogenic species of *Pestalotia*. *Indian Phytopath*., **25** (3) : 410-416.

Ramakrishnan, T.S. 1971 : Diseases of rice. Indian Council of Agricultural Research, New Delhi, 94 pp.

Reddy, E.J.S., Chary, M.P. and Reddy, S.M. 1984 : Evaluation of different fungicides in the control of sheath blight of rice caused by *Acrocylindrium oryzae*. *Indian J. Mycol. Pl. Pathol*., **13** (1) : 105.

Riker, A.J. and Riker, R.S. 1936 : Introduction to Research on plant diseases. John, S. Swift. Co., St. Louis.

Roy, A.K. 1965 : Studies on leaf blight of *Inga dulcis*, *Indian Phytopath*., **18**: 48-49.

Shahjahan, A. K.M., Harahap, Z. and Rush, M.C. 1977 : Sheath rot of rice caused by *Acrocylindrium oryzae* in Louisiana. *Plant Dis. Reptr*., **61**: 307-310.

Sharvelle, E.G. 1960 : The nature and uses of modern fungicides. Burgess Publ. Co, Minn., USA, 308 pp.

Sherf, A.F., Page R.M., Tullis, E.C. and Margan, T.L. 1947 : Studies on factors affecting the infectivity of *Helminthosporium oryzae*. *Phytopathology*., **37**: 281-290.

Singh, N.I. and Devi Tombisana, R.K. 1990 : Survey Report of Botany & Plant Pathology Dept. Manipur Agricultural College, Iroisemba., 5 pp.

Singh, N.I. and Devi, R.K.T. 1991 : Collar rot of rice in Manipur. *Intern. Rice Res. Newsl.*, **16** (1) : 20.

Singh, N.I., Devi, Tombisana R.K. and Nongdrenkhomba S.L. 1995: *Pestalotiopsis versicolor*: A causal agent of rice collar rot. *Plant Disease.*, **79** (11) : 1186.

Sittchai, T. 1974 : Studies on *Rhynchosporium oryzae* Hashioka and Yokogi, the fungus causing leaf scald of rice. *Annual Research Report, Rice Department*, Thailand, 1971, 516-526.

Somogyi, M. 1952 : Notes on sugar determination. *J. Biol. Chem.*, **195**: 19-23.

Srivastava, H.P. 1965 : D. Phil Thesis, University of Allahabad.

Subramanian, C.L. and Ramakrishnan, G. 1975 : Varietal reaction to the sheath rot of rice (*Acrocylindrium oryzae* Saw). *Curr. Sci.*, **44** (11) : 405-406.

Sugha, S.K. and Singh, B.M. 1989 : Effect of glume blight of rice on grain yield and quality. *Indian Phytopath.*, **42**: 163-164.

Sun, S.K. and Snyder, W.C. 1978 : The bakanae disease of rice plant. *Science Bulletin, Taiwan.*, **10** (7) 2, (8) 4, 9 (4), 10 (4).

Swamy. R. R. and Mani, K. 1978 : Soporulation of *Pestalotia palmarum* Cooke in culture Proc-India-Acad Sci. Sec.-B, **87** (5) : 93-98.

Tandon, R. N. and Bilgrami, K. S. 1957 : Cultural studies of *Pestalotia mangiferae* But. with special reference to its sporulation. *Proc. National Academy Science.*, **27**: 96-105.

Tandon, R.N. and Mitra, S.K. 1963 : Some physiological studies on *Pestalotiopsis versicolor* (Speg.) Stey. *Proc. Nat. Acad. Sci.* (India)., **33**: 583-590.

Tasugi, H. and Ikeda, Y. 1956 : Studies on the sheath rot or rice plant caused by *Acrocylindrium oryzae* Sawada. *Bull. Nat.* Agric. *Sci. Tokyo, Ser.* C., **6**: 151-166.

Thakur, K.S.S 1973 : Effect of rufanic acid on the foot-rot disease of rice caused by *Fusarium moniliforme*. *Phytopathologische Zeitschrift.*, **78**: 81-85.

Thind, K.S. and Randhawa, H.S. 1957 : Studies on the nutrition of the fungi II. The influence of carbon sources on the growth and sporulation of *Colletotrichum capsici*. *Proc. Nat. Acad. Sci.* (India), **27**: 47-58.

Thomas, K.M. 1937 : Administrative report of the Government Mycologist for the year 1936-1937. In Administrative Reports: Agriculture, Chemistry, Entomology and Mycology, Madras, 1936-37, 35-51.

Tikoo, M. L. 1985 : Effect of Sheath rot (*Acrocylinderim oryzae*) on some yield components of rice. *Research* and *Development Reporter.*, **2** (2) : 24-25.

Upadhyay, R. K. 1984 : Evaluation of fungicide and antibiotics against *Pestalotiopsis funerea in vitro* causing leaf spot of *Ecualyptus globulus*. Indian Forester., **110** : 209-217.

Venkatasubbaiah, P. 1985 : Influence of soil pH on growth and infectivity of *Rhizoctonia solani*, the incitant of collar rot of coffee seedling *Geobios*, India, **12**: 148-150.

Vidhyasekaran , P. and Lewin, H.D , 1987 : Time of spraying to control sheath rot (ShR). *Intern. Rice Res. Newsl.*, **12** (6): 21-22.

Wang, S.Y.C. and Tourneau, D.L. 1972 : Amino acids as nitrogen sources for growth and sclerotia formation in *Sclerotium. Trans . Brit. Mycol. Soc.*, **59**: 509-512.

Wangsasitorn, D. and Supavita, P. 1974 : Study on varietal reaction of rice collar rot, and host range of *Ascochyta oryzae*. Annual Research Report, Rice Department, Thailand 1971, 510-512; 513-514. (Thai, En).

Wills, C.B. 1968 : Effect of various nitrogen sources on growth of *Sclerotinia., Can. J. Microbiol.* **14**: 1035-1037.

Wooster, R.C. and Cheldelin, V.H. 1945 : Growth requirements of *Penicillium digitatum*. Arch. *Biochem.*, **8**: 311-320.

Yadav, R.N., Shukla, P. and Dwivedi Kusum. 1986 : Detection of amino acid in Corynespora blight of Sesame (*Sesamum indicum* L.) *Indian J. Mycol. Pl. Pathol.*, **16** (2) : 182.

Yamaguchi, T. and Ito, I. 1975 : Spray inoculation with conidial suspension of *Fusarium* leaf spot fungus of rice plant. *Ann. Phytopathol. Soc.* Japan., **41**: 500-501 (Ja. En).

Yu, K.S. and Sun, S.K. 1978 : Influence of nutrition and light on sporulation of the rice bakanae fungus, *Fusarium moniliforme* Sheldon. *Plant Protection Bulletine*, Taiwan., **20**: 141-150 (Ch, En).

Zuge, G.H. and Lapis, D.B. 1985 : Studies on rice sheath rot. I. Pathogenicity of the pathogen and the factors affecting the disease. *Acta Phytopathologica Sinica.*, **15** (1) : 1-8.

INDEX